Sources and Studies
in the History of Mathematics
and Physical Sciences

For further volumes:
http://www.springer.com/series/4142

John M. Steele

Ancient Astronomical Observations and the Study of the Moon's Motion (1691–1757)

 Springer

John M. Steele
Department of Egyptology
and Ancient Western Asian Studies
Brown University
Providence, RI 02912
USA
john_steele@brown.edu

ISBN 978-1-4939-4361-6   ISBN 978-1-4614-2149-8 (eBook)
DOI 10.1007/978-1-4614-2149-8
Springer New York Dordrecht Heidelberg London

Printed on acid-free paper

Springer is part of Springer Science+Business Media (www.springer.com)

To F. Richard Stephenson in honour of his pioneering work in the scientific use of ancient eclipse records

Preface

The end of the seventeenth and the first half of the eighteenth century witnessed a revival of interest in ancient astronomy among astronomers. The investigation of "secular accelerations" of heavenly bodies—gradual changes to their mean velocities—relied upon the comparison of ancient and contemporary observations. These secular accelerations are extremely small, changing the mean velocities of the moon and planets by only arcseconds per century, but their cumulative effect can be detected from even crude observations made over very long timescales. In order to examine the secular accelerations in detail astronomers were forced to use the small number of lunar, solar, and planetary observations made by ancient and medieval astronomers. Knowledge of the secular accelerations was important both for constructing accurate astronomical tables (of particular significance for the moon because of its use in longitude determination) and because the existence of secular accelerations provided a crucial test of the power of the gravitational theory in explaining the motion of heavenly bodies. But the historical astronomical observations available to the seventeenth- and eighteenth-century astronomers were few in number and of doubtful reliability.

Until the middle of the seventeenth century ancient astronomy was still a living tradition, understood in detail by most astronomers, even if its basic theories had been superseded by Kepler's new astronomy. But by the end of the seventeenth century, only 50 years later, ancient astronomy had become a foreign land: its methods all but forgotten except in the most general terms. The astronomers who needed to use ancient astronomical observations to investigate the secular accelerations were faced with the task of extracting the observations from the original texts, understanding how they were observed and the terminology used in recording them, evaluating their reliability, and developing a methodology for their use. This book investigates the use of ancient and medieval observations by astronomers to investigate the moon's secular acceleration during the period from about 1691 to 1757. Although the secular accelerations of Jupiter and Saturn were discovered somewhat earlier than that of the moon, the moon's secular acceleration was the most actively studied during the eighteenth century because of the greater number of ancient observations available, allowing the magnitude of the acceleration to be determined with greater confidence and precision. In a little under a decade starting in 1749, three estimates of the size of the moon's secular acceleration were published: first by Richard Dunthorne, then by Tobias Mayer, and finally by Jérôme Lalande. Mayer and Lalande are fairly well known to historians of science, but the same cannot be said of Dunthorne. He does not appear in the *Dictionary of Scientific Biography*, for example, and only short (and not fully accurate) biographies of him are given in the *Biographical Encyclopedia of Astronomers* and the *Dictionary of National Biography* (in order to rectify the lack of published information on Dunthorne's life, I have included a detailed biographical account in the beginning of chapter 7). Dunthorne, Mayer, and Lalande investigated the secular acceleration using largely the same historical data. They differed, however, in how they analysed the ancient and medieval observations, and, in particular, in the relative trust they placed in the different historical sources.

My aim in this book is twofold. First, I try to tell the story of how the secular acceleration of the moon was discovered, the reception of its discovery, and attempts to determine its size from historical data in the period up to about 1757. After this date, the study of the secular acceleration switched from investigating whether the acceleration existed and determining its size from the ancient data to attempting to account for the acceleration theoretically. Secondly, I attempt to address the wider question of how ancient and medieval astronomy was viewed in the eighteenth century. In particular, I investigate European perceptions of astronomy from different cultures: Ancient Greek, Arabic, Babylonian, and Chinese. It will be clear, for example, that biases against contemporary cultures in the far and near east influenced views of the ancient and medieval astronomy of those regions.

***

Documents from the seventeenth and eighteenth century, both manuscript and print, are remarkable for their inconsistency in the use of personal and other names. Differences in spelling are common, as are the use of Latinized names alongside vernacular forms. These issues are even more apparent with attempts to render Arabic and Chinese into Latin script. Except within direct quotations, I have chosen to use the commonly accepted modern forms of names (for example, Delisle not De l'Isle, al-Battānī not Albategnius). In a few instances I have given both the transcription used at the time and the modern transcription in order to avoid confusion; this is particularly the case with the names of works in Chinese which, as rendered in the eighteenth century, may be hard for the non-specialist reader to identify with the modern Pinyin form (for example, 竹書 was transcribed by Gaubil as "Tchou chou" but is rendered in Pinyin "Zhushu").

Another complication in dealing with the eighteenth century is the use of different calendars in Great Britain and on the continent. Britain adopted the Gregorian calendar in 1752. "Old style" Julian dates were used prior to 1752 in which the year began on 25 March. Dates from 1 January to 24 March were usually accompanied by a double year number, for example 1750/1 meaning that the date was in the last part of the year 1750. Astronomical tables, however, always employed "new style" dates in which the year began on 1 January. I have retained "old style" dates for events in Britain. Dates relating to continental Europe, where the Gregorian calendar was adopted much earlier, and British dates after 1752 are given in the Gregorian calendar.

Unless otherwise noted all translations are my own.

***

This book has its origin in an invitation to speak on the topic of the use of ancient eclipses in early studies of the moon's secular acceleration at the conference "Ptolemy in Perspective" organized by Alexander Jones and hosted by the California Institute of Technology in 2007. That I knew anything about this topic was due to F. Richard Stephenson, who supervised my PhD on ancient eclipse records at the University of Durham during the period 1995–1998. He first introduced me to the problem of the moon's secular acceleration and the variability in the Earth's rate of rotation, and encouraged me to explore the history of astronomy across a variety of cultures. In undertaking the research for this book I have been fortunate to be able to discuss aspects of it with many colleagues including the late John Britton (whose own study of Ptolemy and the history of the secular acceleration provided a framework for my own work), Steven Wepster (who generously pointed me towards the relevant manuscripts of Tobias Mayer), Alexander Jones, Noel Swerdlow, Jed Buchwald, and Len Berggren. I gratefully acknowledge the libraries and archives that have allowed me to consult and discuss books and manuscripts in their collections: the Museum of the History of Science, Oxford; the Niedersächsische Staats- und Universitäts-Bibliothek, Göttingen; the University Library, Cambridge; the British Library; the Crawford Library, Royal Observatory in Edinburgh; the John

Hay Library, Brown University; Palace Green Library, University of Durham; the library of the Royal Society. I also wish to thank William S. Monroe for help reading work in eighteenth-century Dutch. Finally, my biggest thanks go to my wife Rebecca for her support during the writing of this book and her help reading tricky passages in French.

Providence, RI, USA   John M. Steele

Contents

List of Figures

Chapter 1
Introduction

I received yours of the 22^d last and do assure you that I shall endeavour to improve the Opportunity I now have of seeing many Authors, to the advantage of Astronomy, I have been very diligent in collecting all that is valuable among the sacred remains of Antiquity relating to the Moon, wch are only Lunar Eclipses, besides a few Applications of ☽ to fixt Stars, and three Eclipses of the Sun (one observed by Theon Alexandrinus Junior father of the Celebrated Hypatia at Alexandria A.D. 365 and two by Albategnius).

—Letter from Richard Dunthorne to Samuel Rouse dated 13 December 1740 (MS Museum 95, p. 100).

On page 174 of the *Philosophical Transactions* for 1695 Edmond Halley declared that he believed that the moon's motion was gradually accelerating:

And I could then pronounce in what Proportion the *Moon's* Motion does Accelerate; which that it does, I think I can demonstrate, and shall (God willing) one day, make it appear to the Publick.

Halley's announcement came at the end of a paper discussing the ancient city of Palmyra in Syria and the inscriptions that had been found there. After outlining the history of the city from Classical sources, Halley made some remarks on the inscriptions, noting, for example, that before A.D. 500 none of the individuals referred to in the inscription had Roman names whereas after that date forenames taken from Roman Emperors such as Julius, Aurelius and Septimus, become common. He then turned to the geography of the region identifying Aleppo, Andrene and Efree with the ancient cities of Berrhæa, Androna and Seriane respectively and stating that Ptolemy and more recently Kepler in the *Rudolphine Tables* (followed by Bullialdus and others) gave incorrect latitudes and longitudes for these cities. More importantly, Halley said, he had identified the ancient city of Aracta with the city now known as Racca (al-Raqqa) on the Euphrates. It was in this city that al-Battānī made his observations. If an accurate latitude and longitude for this city were known, Halley said, it would be possible to use al-Battānī's observations to resolve the question of whether there has been any change in the axis of the Earth's rotation and to determine the size of the acceleration of the moon's motion.

Since antiquity it had been known that the moon's motion was not constant. The daily motion of the moon as seen against the fixed background of stars varies between about 11° per day and about 15° per day over a cycle of a little over 27½ days known as the anomalistic month. Ptolemy identified two contributions to the moon's variable motion: a first anomaly that is dependent upon celestial longitude (often referred to as "zodiacal anomaly") and a second anomaly dependent upon the

J.M. Steele, *Ancient Astronomical Observations and the Study of the Moon's Motion (1691–1757)*, Sources and Studies in the History of Mathematics and Physical Sciences, DOI 10.1007/978-1-4614-2149-8_1, © Springer Science+Business Media, LLC 2012

distance of the moon from the sun (named "evection" by Bullialdus in 1645). In the sixteenth century Tycho Brahe discovered a third anomaly, called by him the "variation", which caused the moon to speed up around syzygy and slow down around the quadratures. These anomalies all caused the moon to cyclically speed up and slow down in its motion, but the mean motion of the moon remains constant. What Halley had claimed to identify was another type of acceleration in the moon's motion: a long-term change in the mean motion of the moon itself. One consequence of a lunar acceleration of this type is that parameters such as the mean length of the synodic month would not be constant over time.

The idea that fundamental astronomical parameters such as the mean length of the tropical year or the mean length of the synodic month might change on long timescales was not a new one in astronomy. In book 3 of the *Almagest*, for example, Ptolemy discussed Hipparchus's suspicion that the length of the tropical year varied, noting, however, that Hipparchus was unable to confirm this from the observations of solstices and equinoxes available to him. Ptolemy himself concluded that the length of the tropical year is constant, famously deriving the same slightly high value for the length of the tropical year as Hipparchus, using observations that he clearly tampered with to obtain this result.[1] Later, the discrepancy between Hipparchus's (and Ptolemy's) year length and measurements of the year length made by medieval Islamic astronomers led to attempts to explain this change in year length through a model of variable precession.[2] Similar ideas were put forward by Renaissance European astronomers such as Tycho Brahe and Christian Longomontanus.[3] It was always assumed that the mean motion of the sun itself was constant, however. Changes in the length of the tropical year were due to changes in the position of the vernal equinox, not due to a change in the speed of the motion of the sun.

Johannes Kepler appears to have been the first to suggest that the mean motion of a heavenly body may itself change over time.[4] Writing to Matthias Bernegger in June 1625, Kepler announced that "I have learned from the observations of Regiomontanus and Waltherus that all the positions of Saturn in particular, if not also those of Jupiter and Mars, everywhere in the eccentric circles, were then more advanced or less advanced than would be implied by mean motions running uniformly from Ptolemy to Tycho".[5] Kepler continued by remarking that there is need for a "secular equation" (*aequatione seculari*), but that it will be impossible to determine what kind of equation this will be until several more centuries of careful observations have been made. Kepler remarked further on the need for secular equations in his *Rudolphine Tables* of 1627, again citing the observations by Regiomontanus and Walther as evidence, and further remarking that the motions of the sun and moon are also not precisely uniform. He promises that he will write a small book on the subject, but the promise was never fulfilled.

Kepler's remarks were noted by the English astronomer Jeremiah Horrocks who agreed with Kepler that there was evidence for the mean motion of Jupiter and Saturn having changed since Ptolemy's time, and through Horrocks's work by John Flamsteed.[6] Flamsteed grappled with the problem of the inequality in the motions of Jupiter and Saturn throughout his career as Astronomer Royal, without reaching a satisfactory explanation for the cause of the inequality or of how to take account for it in calculating positions over long timescales. What is important for

---

[1] See, for example, Jones (2005). Claims that Ptolemy fudged these observations go back at least to the Renaissance.

[2] Ragep (2010).

[3] Swerdlow (2010).

[4] Wilson (1985).

[5] Kepler's letter was published in *Epistolae J. Keppleri et M. Berneggeri mutae, Argentorati* (1672), pp. 68–75. The translation quoted is taken from Wilson (1985), p. 36.

[6] Wilson (1985), pp. 38–51.

the present discussion, however, is that in the 1690s Flamsteed and Newton, with whom he was still on working terms at the time, asked Halley help make calculations in order to investigate the problem. The idea of secular changes in the motions of heavenly bodies was therefore not completely new—either to Halley or to his audience—when he made his announcement of an acceleration in the moon's motion.

Halley's brief announcement at the end of the 1695 paper on Palmyra was his only published words on the moon's acceleration. He apparently never returned to the topic, despite his later extensive investigations of lunar motion. Gradually, however, other astronomers paid attention to the question of whether there was a secular change in the motion of the moon, sun and planets and by the middle of the eighteenth century it had become an important topic within astronomy. In the 1740s Euler proposed that the Earth's motion around the sun was affected by resistance to the æther. This would cause a secular acceleration in the sun's motion as seen from the Earth and Euler derived the resulting secular equation for correcting the sun's position which he included in his solar tables. From 1749 to 1757 three attempts were made to derive the coefficient of the moon's secular equation. The first, a groundbreaking study by the English astronomer, surveyor and butler at Pembroke Hall, Cambridge, Richard Dunthorne, obtained a value of $10''$ per century^2, essentially the value that would later be taken as correct by theoretical astronomers trying to explain the cause of the acceleration. Four years later the German astronomer Tobias Mayer obtained a much lower value of $6.7''$ per century^2 which he incorporated into his lunar tables. Four years later again, the Frenchman Jérôme Lalande reviewed and reinvestigated the problem, obtaining a value for the moon's acceleration that was essentially in agreement with Dunthorne. Around the same time, astronomers and mathematicians began applying perturbation theory to problems of celestial mechanics, prompted in part by a series of prize competitions set by the Paris and Berlin Academies. After 1757, attention turned away from proving the existence of the moon's secular acceleration and determining its magnitude towards attempting to explain the cause of the acceleration within the framework of gravitation theory. By the last decade of the eighteenth century it appeared that the problem had been solved by Laplace (it would be the mid-nineteenth century before an error in Laplace's argument was uncovered by John Couch Adams).[7]

The discovery and measurement of the secular acceleration of the moon relied on the study of ancient and medieval observations of lunar and solar eclipses. These "sacred remains of Antiquity" as they were called by Richard Dunthorne were few in number—the eclipses described in Ptolemy's *Almagest*, a solar eclipse observed by Theon of Alexandria in the fourth century A.D., and a handful of observations by Islamic astronomers of the medieval period were all that were available—and often less clearly reported than might have been hoped. Nevertheless, as Dunthorne in particular was to show, when used judiciously and imaginatively they provided just enough information to allow the size of the moon's secular acceleration to be constrained.

Astronomy always has been and still is a science that relies on the use of past observations. Unlike most sciences, astronomy can never be truly experimental: astronomers can only observe the astronomical phenomena that present themselves. If, for example, a particular planetary configuration is required to determine a parameter for an astronomical theory, astronomers must either wait until that configuration comes around or make use of observations made at an earlier occurrence of that same event. Investigating the anomalistic motion of Saturn, say, an astronomer may have to wait up to 30 years for Saturn to reach a particular celestial longitude. Furthermore, many astronomical phenomena, such as precession and the proper motion of stars, become detectable only on long time scales, often longer than one person's lifespan. Such phenomena can only be investigated through the use of observations made by previous generations of astronomers. Perhaps uniquely in the sciences, astronomers,

---

[7]Britton (1992), pp. 153–178 provides a useful overview of the history of work on the secular acceleration of the moon.

therefore, are forced to rely upon empirical data collected by their predecessors. Generally when using such data astronomers will need to consider the data's reliability and accuracy which may involve critically analyzing the observational techniques and instruments used to collect the data or simply accepting stated error estimates provided by the originator of the data.

Different issues arise when using observations that were made not just a few years or decades earlier but hundreds of years or millennia ago. These can include the conversion between different chronological systems perhaps leading to uncertainty in the dating of observations, different languages and uncertainty over the precise meaning of terminology, lack of clear distinction between observations and predictions or calculated data, and the different aims of different astronomies where accuracy in observation may not have been of importance. Despite these potential problems there are questions in present day astronomy that can be addressed only through the use of ancient and medieval observations either because the phenomena only occur very infrequently (for example, identifying historical supernovae in our galaxy with their remnants) or because changes only occur over long timescales (for example, gradual changes in the Earth's rate of rotation or variability in solar activity). This field of research has become known as "applied historical astronomy".[8] I have argued elsewhere that what distinguishes applied historical astronomy from the normal use of previous observations by astronomers is a reliance on observations made in a different cultural context from which the astronomer is working. For example, an astronomer using the catalogue of galaxies by Harlow Shapley and Adelaide Ames published in the 1930s knows that the way the data was collected and presented is essentially the same as how it would be today. The catalogue itself may be considered out of date but a user would not face any obstacles in obtaining data from it. By contrast, an applied historical astronomer who uses a Chinese observation of a solar eclipse from the Han dynasty needs to be able to translate the record from Chinese, understand enough of the Chinese calendar both in its operational rules and knowing the sequence and length of reign periods in which years are given to be able to convert the date into the Julian calendar, be able to work out from contemporary historical sources where the observation was likely to have been made, and understand enough of the cultural context of the record to be able to assess its reliability.[9]

There are several examples of applied historical astronomy throughout history. Ptolemy used observations of lunar eclipses and planetary positions made in Babylon from the eighth to the third century B.C., half a millennia before his own time, which he (or an earlier astronomer on whom he relied) had to correlate with dates in the Egyptian calendar and figure out the meaning of foreign astronomical terminology.[10] The thirteenth-century Chinese astronomer Guo Shoujing 郭守敬 used ancient and early medieval observations of solstices, equinoxes and solar and lunar eclipses to test his "Season Granting System" (*shoushili* 授時曆) of calendrical astronomy and likewise had to address various chronological and terminological problems.[11]

By the eighteenth century, Ancient Greek-style geometrical astronomy, which had formed the basis for all astronomy in the western world from Ptolemy, through the medieval Islamic astronomers into the astronomy of Renaissance Europe, was no longer a living tradition. Astronomers of the sixteenth and early seventeenth century, such as Tycho Brahe, Kepler, Riccioli and Bullialdus, had a deep and full understanding of Ptolemy's astronomy. The epicycle, eccentric and the equant may have been rejected by Kepler and replaced with elliptical orbits, but both he and the next generation of astronomers

---

[8] Stephenson and Clark (1978), Steele (2004).

[9] Addressing these problems can be quite difficult and sadly is not always done with the care and rigour required—see my comments in Steele (2004). In Steele (2003) I discuss related problems in the context of attempts to establish absolute chronologies using astronomy. For a detailed discussion of how these problems can be successfully addressed, see Stephenson (1997) which includes many examples.

[10] Steele (2004), Jones (2006), Steele (2011).

[11] A translation and study of Guo Shoujing's work is given in Sivin (2009).

understood their mathematics and why and how Ptolemy used them. This understanding went beyond the superficial comprehension that might have been needed in order to argue against Ptolemy's models. These astronomers had been brought up in the Ptolemaic tradition and, as Kepler's famous remark about Copernicus modeling Ptolemy not reality exemplifies, at least in part approached their understanding of the new astronomy from a Ptolemaic viewpoint. For several decades after the publication of Kepler's *Astronomia Nova* the authors of astronomical books debated the reality of his system, problematical to some because of its basic assumption of heliocentricity. Several early to mid-seventeenth-century books present both Kepler's elliptical orbits and a geocentric model (usually Tycho's Earth-centered system in which the sun moves around the Earth but the other planets travel around the sun) incorporating epicycles and eccentric circles. These debates had ceased by the end of the seventeenth century, however. Kepler's laws had by then become more or less universally accepted, thanks in part to refinements in the way they were applied by Horrocks and others and, most significantly, a physical argument in support of elliptical orbits coming out of Newton's gravitational theory. Eighteenth-century astronomers no longer had any need to consider planetary models in the Ptolemaic tradition, and within a very short time, understanding of the theories of epicycles, eccentric circles and the equant was lost. As I shall discuss in Chap. 5, early eighteenth-century histories of astronomy show almost no interest and very little comprehension of ancient and medieval astronomical theory. For the authors of these works, what was of importance was telling a story of the progression of astronomy from crude observations made by the ancients to the accurate instruments and observations of their own time. The eighteenth-century astronomers who used ancient and medieval observations to determine the secular acceleration of the moon may have been able to read Greek and Latin sources with ease, but they would have had almost no background in the practice of ancient astronomy itself. For them, in a way that was not true a century earlier, the astronomy of the ancients was astronomy from a different culture.

The Uses of Antiquity

Ancient and medieval sources provided the observations necessary to investigate several problems within astronomy in addition to the moon's secular acceleration, including determining the orbits of comets, the change in the obliquity of the ecliptic, the proper motion of stars and the mean motions of the sun, moon and planets. But providing data for the determination of astronomical parameters was not the only use of antiquity by eighteenth-century astronomers. The history of astronomy was also invoked for a variety of pedagogical and rhetorical uses, of which I will briefly outline below two of the most common: providing examples for the presentation of basic astronomical ideas in introductory texts of astronomy, and being invoked in support of assertions within current astronomical arguments. The most frequent use made of ancient astronomy, however, was in the study of chronology. Ancient astronomical texts provided the main source of information about ancient calendars, and ancient observations were investigated to try to establish absolute chronologies of the classical and biblical world.

Many eighteenth-century authors published introductory books on the subject of astronomy. These included both professional astronomers and mathematicians such as David Gregory, the Savilian Professor of Astronomy at Oxford, William Whiston, Newton's successor at Cambridge, and Jérôme Lalande, who held the chair of astronomy at the Collège de France, and amateurs including Charles Leadbetter, James Ferguson and George Costard. Roger Long's *Astronomy, in Five Books* is typical of such works and provides a convenient example of how history was used in the presentation of astronomical concepts. Long (1680–1770) was Master of Pembroke Hall and from 1750 Lowndean Professor of Astronomy and Geometry at Cambridge. Long had entered Pembroke as a student in 1697 and was elected to a Fellowship in 1703. Except for a short period as a tutor to the family of

Sir Wolstan Dixie in Market Bosworth, Long spent most of his long life at Pembroke, becoming Master in 1733 at the age of 50, and holding that office until his death in 1770. He was also Rector of Orton Wellville in Huntingdonshire from 1716–1751, and Vicar of Cherry Hinton in Cambridgeshire in 1729, that same year being elected a Fellow of the Royal Society. Long had wide interests ranging from astronomy, mathematics and the construction of ingenious devices such as the "water-work" in his garden in which he could peddle around on a water-cycle, to music, scripture (he took his DD in 1728) and university politics (he held the office of Vice-Chancellor of the University in 1733, stood as a candidate for the High-Stewardship of the University in 1764, and unsuccessfully stood again for Vice-Chancellor in 1769, at the age of 89).[12] Long's *Astronomy* was published serially, the first part appearing in 1742, the second in 1764 and the last part (completed by Richard Dunthorne and William Wales after Long's death in 1770) appearing posthumously in 1784.[13]

Long was never an active research astronomer but he seems at least to have been more dedicated to teaching than most Cambridge professors of the day (there is evidence that Long was still giving astronomical lectures as late as November 1768).[14] His *Astronomy* almost certainly came out of his teaching. Over its five books, Long presented a complete survey of astronomy ranging from basic spherics, the Copernican model of the solar system, and the distance and movement of the stars to the moon and its motion, the planets and comets. He also devoted considerable attention to the philosophical (meaning Newtonian) basis of current astronomy, units of measurement and, in book five, provided a fairly detailed history of astronomy from earliest man to his own day (most of the fifth book was in fact written by Dunthorne and Wales after Long's death). Long's presentation is sparing in its use of mathematics—a subject on which Richard Dunthorne felt Long was largely ignorant[15]— but very clearly written compared to the more mathematical books by Gregory and Ferguson which covered most of the same topics.

On several occasions through his book Long used examples from ancient Greek astronomy in his presentation of aspects of basic astronomy. For example, discussing the calculation of the circumference of the Earth, Long described Eratosthenes's well-known solution to the problem.[16] Similarly, in describing how to determine the horizontal parallax of the moon, Long summarized Ptolemy's method from *Almagest* V.13. After reporting Ptolemy's result, Long remarked "This is manifestly too great by several minutes, and is mentioned here only to shew the method".[17] Ptolemy was used again in Long's discussion of how to find the diameter of the Earth's shadow where he gave a short account of Ptolemy's method from *Almagest* V.14.[18] In both these cases, however, the method Long presented is far simpler than what Ptolemy actually wrote and Long silently inserted his own terminology and cited his own propositions in the account. Long's Ptolemy is an idealized Ptolemy, a Ptolemy whose astronomy is simple to understand and clean, without any of the fine detail (and complications) of the real Ptolemy. The same simplifications can be found in Long's discussion of Ptolemy's planetary theory, which qualitatively describes epicycles and eccentrics, but does not mention the equant.

Similar simplifications of Ptolemy (and even more so Copernicus) can be found in many introductory textbooks of astronomy today. Their aim, like that of Long, is to provide a simple context to explain some rather boring aspects of technical astronomy. The use of historical figures— however, anachronistically—allow the authors to justify the presentation of a simplified model

---

[12] Attwater (1936), pp. 91–101.

[13] Jungnickel and McCormmach (1996), p. 124.

[14] *The Cambridge Chronicle* 26 November 1768; see Winstanley (1935), p. 172.

[15] MS Museum 95, p. 70.

[16] Long, *Astronomy*, p. 127.

[17] Long, *Astronomy*, p. 371.

[18] Long, *Astronomy*, p. 385.

where the true model would be too difficult to understand. Using a historical figure—and everyone knows that people in the past did not understand things like we do today—is meant to indicate to the student that the model is just an approximation. Long and other eighteenth-century authors seem to have used the names of ancient Greek astronomers for the same purpose.

A second use of antiquity by eighteenth-century astronomers was the invocation of ancient astronomers for rhetorical purposes in the course of astronomical arguments. Ancient authors, especially well-known names, could be used either in support of a line of reasoning, providing the argument with authority and history, or as straw men who could be argued against in order to imply that the alternate view being put forward is much more reasonable. Such practices have a history stretching back into antiquity itself, and work particularly well when the name of the astronomer is familiar but his astronomy is less well known. Berossos, for example, was used in this fashion by Cleomedes and Vitruvius, both of whom describe a model for the illumination of the moon which they attribute to him and follow with arguments for an alternative model that they see as much more plausible. Berossos was useful since his name carried connotations of great learning from the east, and if his model was less satisfactory than an alternative, then that alternative, and the person who developed it, must be very good indeed.[19] In the sixteenth century Copernicus cited several ancient authors in the introduction to *De Revolutionibus* to show that although it was commonly held that the Earth was immobile he was not the first or only person to consider the possibility that it moved. In contrast to the ancient uses of Berossos, Copernicus cited these ancient figures, about who almost nothing is known, in support of his own arguments. By showing that his claim that the Earth moved had historical precedents he could deflect criticism that his idea was totally new and unfounded and situate himself within the tradition of western astronomy.

Similar uses of antiquity can sometimes be found in the works of late seventeenth- and early eighteenth-century scientists. Halley, for example, when presenting his claim that the Saros cycle of 223 lunar months can be of use in examining the moon's motion and in predicting eclipses, took pains to trace the lineage of the Saros back through classical sources to the Babylonians, even incorrectly reconstructing its name (see Chap. 2). Newton supported some of his arguments about the role of God in the universe given in the general scholium of the *Principia* by making reference to Pythagoras, Thales, Anaxagoras and Aratus among others.[20] And, slightly earlier, Hooke drew on ancient sources and authority in his presentation of the theory of the Earth.[21] Antiquity may not have held the same authority that it had possessed during the Renaissance, but it could still be invoked to buttress an argument.

The most common application of ancient astronomical sources in the seventeenth and eighteenth century was to the study of chronology. Already in the sixteenth century, astronomy was recognized as the most important weapon in a chronologist's armoury. Ancient astronomical texts provided the most extensive accounts of ancient calendrical practices available to a chronologist and Ptolemy's list of kings' reigns provided the framework on which classical and near eastern chronology could be built.[22] But in addition, ancient reports of astronomical phenomena which could be calculated using current astronomical theories could potentially be dated by calculating all possible phenomena within a historically plausible window of time. The date of the record would be the calculated phenomenon which provided the best fit with the reported observation or, more commonly, an accumulation of best-fitting phenomena for a group of reports linked by a relative chronology for, as chronologists found, constructing an absolute chronology involved in assembling a large, interconnected patchwork

---

[19] Steele (2012).

[20] See Cohen and Whitman (1999), pp. 941–942.

[21] Birkett and Oldroyd (1991).

[22] On the history of Ptolemy's royal canon and its use by chronologists, see Depuydt (1995).

of dated phenomena that provided a coherent chronological system. Scholars such as Heinrich Bünting, Johannes Kepler and Joseph Scaliger calculated eclipses and other astronomical phenomena and dated many ancient records from classical sources.[23]

Many seventeenth- and eighteenth-century astronomers devoted considerable time to the study of chronology. The focus of these studies was generally to attempt to refine and harmonize biblical chronology, but problems of ancient Greek chronology, which had been debated incessantly for over a century, also attracted considerable interest. For example, in the early 1750s George Costard and William Stukely proposed rival dates for the eclipse supposedly predicted by Thales, and as I will discuss in Chap. 3, William Whiston and others fought over the date of the crucifixion with reference to an eclipse reported by Phlegon. In the late seventeenth century, reports of extremely ancient Chinese observations of eclipses reached Europe and started a flurry of studies attempting either to date the eclipses or to disprove their authenticity by Flamsteed, Whiston, Mayer, Costard and many others.

The study of the secular acceleration of the moon was only one of the uses ancient astronomy was put to by seventeenth- and eighteenth-century astronomers. Consideration of these other uses of antiquity will at times be of help in understanding how the ancient observations were used in investigating the secular acceleration.

Late Seventeenth- and Early Eighteenth-Century Lunar Theory

In order to discuss the ways in which ancient astronomical records were used to investigate the moon's secular acceleration, it will be helpful to outline here the methods and development of the lunar theory during the late seventeenth and early eighteenth centuries.[24] Kepler's discovery of the elliptical orbits of the planets had revolutionized planetary theory but he had been unable to obtain similar success in their application to the moon's motion. The multiple inequalities in the moon's motion (the two first and second anomalies known since antiquity plus the variation discovered by Tycho, the inequality of the nodes and inclination of the moon's orbit also discovered by Tycho, and the annual equation found by Tycho and Kepler independently) meant that a simple ellipse could not be used to model the moon's motion. Instead, the empty focus of the ellipse itself required a motion, and various models in which the focus moved in a circular path were proposed by Bullialdus, Thomas Streete and Vincent Wing.[25] Wing and Streete produced competing sets of lunar tables that were extensively used in the second half of the seventeenth century: Wing's *Astronomica Britannica* of 1652 and Streete's *Astronomia Carolina* of 1661. Both sets of tables were used to produce numerous astronomical almanacs (Wing was himself a prolific compiler of almanacs).[26]

Significant improvements in the lunar theory were made by Jeremiah Horrocks during the 1630s, but were not widely known until the 1670s. Horrocks was born around 1619 in Toxteth near Liverpool. He entered Emmanuel College, Cambridge in 1632, remaining there for 3 years before returning to Toxteth. He was ordained in 1639 and appointed curate at Hoole, but died in January 1641/2.[27] Horrocks's work on lunar theory had remained unpublished until many years after his death

---

[23] Grafton (1993) and (2003).

[24] My discussion is of necessity brief. For detailed overviews with references to further work, see Wilson (1989a, b), (1995), (2010), pp. 9–30, Forbes and Wilson (1995), Linton (2004), and Morando (1995).

[25] Wilson (1989a).

[26] Kelly (1991).

[27] Plummer (1940–1941).

when it was included in his *Opera Posthuma* accompanied by tables prepared (in part inaccurately) by Flamsteed on the basis of the Horrocks's theory. Horrocks modeled the moon's motion using an elliptical orbit with a varying eccentricity and a librating apogee.[28] Horrocks's theory was significantly more accurate than the competing theories of Bullialdus, Streete or Wing—Flamsteed claimed it had an accuracy of better than 2′ compared to errors of about 15′ in the theories of Bullialdus and Streete[29]—and it quickly became widely accepted among British astronomers.

Horrocks's lunar theory formed the basis for Isaac Newton's study of the moon's motion. Newton initially hoped to construct his lunar theory purely on the basis of his theory of gravitation, but, following many failures, he eventually opted instead to build upon Horrocks's work. Thus Newton adopted the idea of an ellipse whose eccentricity and apogee are continually varying, adding several small sinusoidal oscillations in order to achieve a better fit with observation.[30] The resulting lunar theory contained seven equations to be applied as corrections to the moon's mean motion. Newton's first published his lunar theory in a small pamphlet in English with the title *Theory of the Moon's Motion*.[31] The "theory" contained in the pamphlet, however, is only a set of precepts for calculating the lunar position, not a fully justified theory. Newton did attempt to present such a theory based upon the law of gravitation in the second edition of the *Principia*, but had to fudge some of the mathematics to produce a theory that appeared to work.[32]

Newton's *Theory of the Moon's Motion* (i.e. the precepts for calculating the moon's position) was quickly translated into Latin and published by David Gregory in his *Astronomiæ Physicæ & Geometricæ Elementa* of 1702, and then re-translated back into English for the English version of Gregory's book (*The Elements of Astronomy, Physical and Geometrical*, 1715). Several sets of lunar tables based upon Newton's *Theory* were compiled by Wright, Leadbetter, Dunthorne and others during the first half of the eighteenth century.[33]

The next significant step in the development of the lunar theory was the realization that instead of treating the three-body Earth–Moon–Sun problem geometrically it could be investigated algebraically using Leibnizian calculus. This approach relied upon techniques for differentiating and integrating trigonometrical functions developed by Leonard Euler. These techniques allowed the moon's orbit to be considered as a sum of a simple elliptical orbit due to the Earth's gravitational attraction with perturbations due to the sun's gravity (in due course, further perturbational effects due to the gravitational attraction of the planets could also be taken into account). Euler himself was the first to attempt to derive the lunar inequalities using trigonometrical calculus. His lunar tables of 1744 (published, probably with revisions, in his *Opuscula varii argumenti* in 1746) were the first tables to incorporate inequalities due to the sun which were derived directly using inverse square law.[34] His work was quickly followed by that of Alexis-Claude Clairaut, who had used similar techniques to derive the inequalities in the motion of the sun as seen from the Earth and was able to make further refinements to the lunar theory, and by Jean le Rond d'Alembert.

Euler, Clairaut and other mathematicians succeeded in deriving a rigorous lunar theory based upon the inverse square law of gravitation, but they do not appear to have had the skill—or the motivation—to fine-tune the theory using observations. This task was undertaken by Tobias Mayer who, starting from Newton's lunar theory, used his own observations to refine the lunar parameters in order

---

[28] Gaythorpe (1957), Wilson (1987) and (1989a), pp. 197–201.

[29] Linton (2004), p. 234.

[30] Linton (2004), p. 276, Whiteside (1976), Kollerstrom (2000).

[31] Cohen (1975).

[32] Linton (2004), p. 277.

[33] Waff (1977), Kollerstrom (2000), p. 208.

[34] Wilson (1995), p. 91.

to construct lunar tables of significantly greater accuracy than any of its predecessors or contemporaries.[35] Mayer claimed he had tested his tables against 200 observations and found errors of less than 0.5′ in about 90% of them, with no errors in excess of 2′ among the remaining 10%, much less than the errors of 4′ or 5′ in the tables of Euler or Clairaut.[36] Mayer continued to improve his lunar tables until his early death in 1762. His final tables were adopted by Nevil Maskelyne for use in the *Nautical Almanac*.

The Secular Acceleration of the Moon

The secular acceleration of the mean motion of the moon discovered by Halley is now known to be a summation of a real acceleration in the moon's motion and an apparent acceleration due by the gradual slowing down of the Earth's rate of rotation causing a change in the length of the day as a unit of time. Both factors are primarily due to tidal friction in the Earth–Moon system. As the Earth rotates, the tidal bulge in the oceans is carried eastwards. Gravitational attraction between this bulge and the moon transfers energy to the moon's forward motion, with the consequence that the moon rises into a higher orbit. Because the moon is now in a higher orbit, its mean motion decreases. The transfer of energy to the moon also has the effect of slowing the earth's rate of rotation. In addition to the effect of tidal friction, the moon's motion is subject to perturbations from the sun and the planets. The earth's rate of rotation is also affected by post-glacial uplift, core–mantle coupling in the Earth's interior, changing sea levels, etc.

Several authors in the eighteenth century raised the possibility that the moon's secular acceleration might be partly or even wholly an apparent effect due to long-term change in the length of day (see Chap. 9), but observationally the change in the mean motion of the moon was always treated as a simple acceleration. For most purposes, it was not the acceleration itself that was needed, but the effect of the acceleration on the longitude of the moon at a given time. For a secular acceleration α the moon's mean velocity v will be given by $v = v_{epoch} + \alpha t$, where t is the number of years since the epoch (because the moon's secular acceleration is quite small, it is common to give α and t in centuries rather than years). The effect of this secular acceleration on the longitude of the moon will therefore be to add a correction of $\frac{1}{2}\alpha t^2$ to the longitude calculated assuming constant mean motion. This correction, $\frac{1}{2}\alpha t^2$, is correctly known as the "secular equation", and the coefficient $\frac{1}{2}\alpha = c$ is correctly the coefficient of the secular equation. However, it has been fairly common among some authors from the late eighteenth up to the twentieth century to refer to c as the "secular acceleration", when this term should correctly be applied to $\alpha = 2c$. I will always use the term "secular equation" to refer to the correction to the longitude and reserve "secular acceleration" for the change in the moon's mean velocity.

---

[35] Wepster (2010).

[36] Forbes and Wilson (1995), p. 65.

Chapter 2
Edmond Halley's Discovery of the Secular Acceleration of the Moon

And by combining the eclipse observations of the Babylonians with Albategeni's and with today's, our Halley showed that the mean motion of the Moon when compared with the diurnal motion of the Earth, gradually accelerates, as far as I know the first of all to have discovered it.

—Isaac Newton, *Principia*, 2nd ed.
(Cambridge, 1713), p. 481.

The secular acceleration of the moon's motion was discovered by Edmond Halley in the first half of the 1690s. Halley's discovery came about through a combination of factors: a long-standing interest in lunar theory, a willingness to engage with historical sources, and a perceived need to prove that the universe was not eternal.

Edmond Halley

Edmond Halley was born in 1656 at Haggerston outside of London and was educated at St Paul's School, London and Queen's College, Oxford.[1] He was elected a Fellow of the Royal Society in 1678. The following year Halley published a catalogue of the stars of the southern sky (the result of a year observing on the island of St Helena off the coast of South-West Africa). He subsequently visited Hevelius in Danzig and Cassini in Paris, two of the best astronomical observers in Europe at the time, and on his return to London began his own programme of astronomical observation. Probably soon after his return, Halley met Isaac Newton and would become instrumental in the publication of Newton's *Principia* in 1687.

In 1686 Halley was appointed Clerk to the Royal Society and in this capacity oversaw the recording of the *Journal Book* and the publication of the *Philosophical Transactions*. He was very active at the meetings of the Society over the next decade (the *Journal Book* records 146 contributions to their

---

[1] For detailed biographical studies of Halley, see Cook (1998) and Armitage (1966). Further useful information may be found in MacPike (1932) and (1937).

J.M. Steele, *Ancient Astronomical Observations and the Study of the Moon's Motion (1691–1757)*,
Sources and Studies in the History of Mathematics and Physical Sciences,
DOI 10.1007/978-1-4614-2149-8_2, © Springer Science+Business Media, LLC 2012

meetings between January 1687/8 and July 1696),[2] speaking on topics as wide-ranging as magnetic needles, diving bells, natural history, the history of ancient Britain, medical curiosities and remedies, and the application of statistics to calculating life annuity rates. His astronomical contributions included suggestions for the improvement in the design and use of telescopes, lunar theory, and his claim that comets of 1607 and 1682 were the same object.

In 1691 Edward Bernard resigned as Savilian Professor of Astronomy at Oxford and Halley declared himself a candidate, with the support of the Royal Society.[3] The professorship, however, went to David Gregory, who had previously held a chair at Edinburgh. Writing more than fifty years later, William Whiston, Newton's successor as Lucasian Professor at Cambridge who was himself removed from his chair on account of his Arian views, claimed Halley was not elected because of his religious unorthodoxy.[4] As we will see, the suspicion cast on Halley's religious views provided a crucial backdrop to his discovery of the moon's secular acceleration.

Halley remained active within the Royal Society until 1696 when he was appointed deputy to Newton at the Mint in Chester. After 2 years at the Mint, Halley spent the end of the 1690s and the beginning of the 1700s undertaking several sea journeys until standing as a candidate for the vacant Savilian Chair of Geometry at Oxford. This time the election went smoothly: Halley's work for the government undertaking naval surveys counted strongly in his favour, and under the new queen there was less concern over his religious or political views.[5] Halley remained at Oxford until the death of Flamsteed on 31 December 1719; Halley was appointed his successor as Astronomer Royal on 10 January 1719/20 and remained at the Royal Observatory until his own death on 14 January 1741/2 and was buried beside his wife at St Margaret's, Lee.[6]

Halley's Discovery of the Secular Acceleration of the Moon

Three aspects of Halley's researches during the early 1690s came together to lead him to discover the secular acceleration in the moon's motion. First was Halley's work on celestial mechanics, in particular his long-standing interest in lunar theory. Between 1679 and 1684 Halley made numerous observations of the moon's position from London, Paris, and his house in Islington.[7] Based upon these observations Halley had succeeded in obtaining new results for the variation which were published by Newton in the first edition (1687) of the *Principia* Book III, Prop. 29. Following this publication Halley urged Newton to work on perfecting the lunar theory, which he considered to be of great importance because of its use in navigation.[8] Halley's view that the lunar theory needed reforming was recorded by the mathematician and astrologer Henry Coles in his notebook:

> Mr Halley says
> Our lunar theory doth yet want emendation … her motion is perplexed with various inequalities, and those arising from divers causes, the sum of all wch may indeed appear in the visible places of the moon. But it is very

[2]MacPike (1932), pp. 210–238.

[3]Cook (1998), pp. 245–249.

[4]Armitage (1966), pp. 122–123. Cook (1998), pp. 246–249 demonstrates that Whiston's claim is unlikely to be the full story.

[5]Cook (1998), p. 322.

[6]Cook (1998), p. 403.

[7]Cook (1998), p. 356.

[8]Halley to Newton 5 July 1687; edition in Turnbull (1960), p. 482.

difficult to determine them separately, so that concerning the quantity of any Equation the Choisist Astronomers have not as yet agreed amongst themselves.[9]

On 16 October 1689 Halley presented a paper at the Royal Society in which he noted that after 18 years and 11 days "the whole course of the Moon is regular".[10] Halley argued that this period could be used to predict eclipses and presented several series of predicted eclipses to the Royal Society on 9 November 1692. The period of 18 years 11 days, corresponding to 223 synodic months, was well known from antiquity (for example, Ptolemy and Pliny both mention it) and had already been discussed by a number of seventeenth-century authors including Bullialdus who called the period the "Chaldean Period".[11] Halley, however, argued on the basis of a misunderstanding of a passage in Suidas and a correction to Pliny,[12] that period was called the "Saros" by the Babylonians and despite almost immediate criticism of the historical basis for Halley's claims over this name it is still in use today.[13] During his tenure as Astronomer Royal, Halley returned to the Saros, succeeding in making measurements of the moon's position over a complete 18-year period. His notebooks from this period include many comparisons of lunar positions separated by one or two Saroi.[14]

During the 1680s and 1690s Halley also devoted considerable time to the study of antiquity. For example, in 1690 Halley argued that Julius Caesar's invasion of England began on 26 August 55 B.C., and that his legions had landed near the town of Deal in the Downs.[15] By means of a careful study of the classical accounts, Halley reconstructed the date of the invasion, making recourse to an eclipse which fixed the date of death of Augustus and calculations of the dates of full moon in 55 B.C. Having now the date of the invasion and the phase of the moon, Halley investigated the tides in the English Channel to figure out the path of Caesar's fleet and the site of the landing, remarking on "the great use of *Astronomical Calculation* for fixing and ascertaining the Times of Memorial Actions".[16] Around this time Halley also investigated Roman Imperial surveys, making comparisons with his own measurements of Roman roads in Britain, and with modern maps (which often did not stand up to the test).

Halley's interest in antiquity extended to scientific data. He investigated historical weather patterns, population records, and the geographical coordinates of cities and whether there had been any change in their latitude since antiquity (which might be explained by a shift in the Earth's axis of rotation, an idea which Halley rejected). Within his astronomical work, Halley's discovery of the proper motion of stars came out of his re-examination of the stellar data found in Ptolemy, and his work on comets incorporated detailed analysis of accounts of earlier observations. Chapman has argued that Halley was more historically aware than most scientists of the seventeenth and eighteenth centuries, and that this awareness informed Halley's scientific work not only through his knowledge of earlier scientific ideas and observations but also in his very approach to investigating nature.[17] According to Chapman, Halley's philosophy of science was built around the idea that the processes of nature are developmental and have their own history.

The third important part of Halley's research during the late 1680s and the early 1690s was his study and emendation of the Plato of Tivoli's Latin translation of al-Battānī's (Albategnius)

---

[9] BL Slone MS. 2281, f. 64.

[10] MacPike (1932), p. 216.

[11] Bullialdus, *Astronomia Philolaica*.

[12] Halley, "Emendationes … Naturalis Historiae C. Plinii".

[13] For a summary of criticisms of Halley's naming of the Saros, which began in the mid-eighteenth century see Neugebauer (1957), pp. 141–142. Lynn (1889) reports the first evidence from an Assyriologist (Bertin) that "Saros" is not a Babylonian term.

[14] MS RGO 2.6 and 2.8.

[15] Halley, "A Discourse tending to prove at what time and place, Julius Caesar made his first decent upon Britain". See also Cook (1998), pp. 200–201, Chapman (1994), p. 171.

[16] Halley, "A Discourse tending to prove at what time and place, Julius Caesar made his first decent upon Britain".

[17] Chapman (1994).

astronomical work *Zīj al-Ṣābi'*. Two editions of this translation had been published (Nuremberg, 1537, and Bologna, 1645) with the title *De Scientia Stellarum*. Unfortunately, these editions were marred by typographical errors and Halley remarked that Plato of Tivoli was "neither sufficiently versed in the language nor instructed in the discipline of astronomy".[18] Halley's aim was to correct the errors in the reports of observations in *De Scientia Stellarum* and to use them to restore al-Battānī's solar and lunar tables (given in his *Zīj* but omitted from the Latin translation).[19] The observations were the time of the autumnal equinox of A.D. 882 and four reports of eclipses. The eclipses were observed by al-Battānī in al-Raqqa (a lunar eclipse on 23 July 883 and a solar eclipse on 8 August 891) and Antioch (a solar eclipse on 23 January 901 and a lunar eclipse on 2 August 901) and were compared by al-Battānī with computations of the circumstances of the same eclipses using Ptolemy's tables.[20] Halley successfully used the equinox observation to determine the apogee and mean motion of the Sun, obtaining results too small by 42″ compared with al-Battānī's tables preserved in the Arabic manuscript of his *zīj*.[21] He had greater problems establishing the lunar tables, in part due to the errors in the printed text of Plato's translation, obtaining values for the mean Moon too small by about 9′.[22]

Halley's interests in these three areas provided the background for his discovery of the moon's secular acceleration. The catalyst that led to the discovery itself, however, lay within debates over the relationship between scientific reason and scriptural authority. Following the Glorious Revolution of 1688 attitudes against the idea that reason took precedence in understanding nature had hardened, and Halley found himself on the wrong side of the debate. By the beginning of the 1690s stories alleging that Halley had been investigated for atheism were in circulation within the Anglican establishment. As Schaffer has argued, whether or not these reports were true, they were taken seriously, not least by Halley himself.[23] In 1691 when the Savilian Chair of Astronomy at Oxford became vacant, Halley applied but lost the position to David Gregory in part, it seems, because of perceptions of Halley's religious views. One charge against Halley was that he believed in the eternity of the world, a belief at odds with the biblical account of God's creation of the universe. Before the decision had been made, Halley wrote a letter to Abraham Hill dated 22 June 1691 asking him to intercede with the Archbishop of Canterbury, who together with the Bishop of Worcester held control of the appointment. Halley sought time, even "if it be but for a fortnight … (t)his time will give me an opportunity to clear myself … there being a caveat entered against me, till I can shew that I am not guilty of asserting the eternity of the world".[24] For Halley, however, it was not sufficient to simply deny the eternity of the world. He wished to use science to *prove* that the world had a beginning and would come to an end. Rather than simply accept that the world would come to an end on the basis of scripture, Halley used science to give plausibility in support of the Christian position. As argued by Schaffer, Halley here missed the point as his critics' objection was to his very use of physical arguments in theology.[25]

---

[18] Halley, "Emendationes ac Notæ in vetustas *Albatênii* Observationes Astronomicas", p. 913.

[19] Mercier (1994), pp. 192–196.

[20] English translations of these eclipse reports from al-Battānī's Arabic are given in Said and Stephenson (1997), pp. 43–45. In the case of the two eclipses al-Battānī observed in Antioch, he also recorded in his *zīj* simultaneous observations of the eclipses by "someone on our behalf" in al-Raqqa in order to determine the difference of longitude between the two cities. See also Said and Stephenson (1996).

[21] Mercier (1994), p. 193.

[22] Mercier (1994), p. 194.

[23] Schaffer (1977).

[24] See MacPike (1932), p. 88.

[25] Schaffer (1977), p. 28.

In a paper published in the *Philosophical Transactions* in January 1691/2 on the subject of the variation of magnetical needles and presenting his hypothesis of a hollow Earth,[26] Halley remarked that "I think I can demonstrate that the Opposition of the Ether to the Motions of the Planets in long time becomes sensible".[27] A consequence of the retardation of the motion would be a slowing down of the motion of the Earth until it eventually came to rest at some point in the future, implying that the Earth was not eternal. Halley discussed this latter effect more thoroughly in a paper read at the Royal Society on 19 October 1692. The *Journal Book of the Royal Society* contains the following description of Halley's paper:

> October 19, 1692. Halley read a paper, wherein he endeavoured to prove that the opposition of the Medium of the Æther to the Planets passing through it, did in time become sensible. That to reconcile this retardation of the Motions the Ancients and Moderns had been forced to alter the differences of the Meridians preposterously. That Babylon was made more westerly than it ought by near half an hour, both by Ptolomaeus, and those since him. And to reconcile the Observations made by Albategnius at Antioch, and Aracla on the Euphrates, they have been forced to make these places ten degrees more Easterly, than they ought, particularly Mr. Street has made Antioch of Syria in his Table of Longitudes, and Latitudes of places half an hour more Easterly than Babylon, whereas in truth it is about 40 minutes more Westerly. That this difference is found by 4 Eclipses observ'd about the year 900 and that by an Artist not capable of mistaking, that they all 4 agree in the same result and are noe other ways to be reconciled. Hence he argued, that the Motions being retarded must necessarily conclude a finall period and that the eternity of the World was hence to be demonstrated impossible. He was ordered to prosecute this Notion, and to publish a discourse about it.[28]

Halley never published his paper but his manuscript of the lecture is preserved in the Royal Society archives.[29] The paper, entitled "Concerning the motion of light" draws upon Rømer's work on the velocity of light to conclude that the æther is about eight million times more rare than our air. Rare as this is, the resistance caused by the æther on the motions of the planets is nevertheless sensible:

> Now if we come to consider how great a quantity of this Æthereal matter they (the planets) penetrate and with how great a velocity it will not withstanding its great subtility seem reasonable that some part of their motion should be taken off by the opposition of this medium, which tho' it be to be expected but a very small matter yet in Multitudes of years it ought to become sensible. This is what I think to have discovered by a long carefull comparison with all that antiquity has left us relating to the Sun and the Moons motion, and I doubt not but to make it appear that the length of the year grows longer and longer and that in that supposition it will be impossible to reconcile the undenied observations of the Ancients with the curious accounts we have of these motions from Tycho Brahe's time onwards.[30]

Halley claimed that the apparent motion of the sun was decelerating, in accordance with what would be expected from the resistance to the passage of a body (the Earth) through the ether, and that earlier astronomers, both ancient and modern, unaware of this deceleration had to shift the meridians of places where observations were made in order to reconcile earlier observation with those of their own time. He cited the discrepancy between the longitudes of Babylonian given by Ptolemy in the *Almagest* (which he believed have been adjusted to make the observations fit) and the *Geography* (where the true longitude is given) as ancient evidence of this practice, and longitudes in Thomas Streete's *Astronomia Carolina* as a recent example.

Halley then turned to the reports of ancient and medieval eclipses recorded in the *Almagest* and by al-Battānī, these being the most ancient ones we have "excepting the fabulous ones of the Chinese".

---

[26] On Halley's hollow Earth theory, see Kollerstrom (1992).

[27] Halley, "An Account of the cause of the Variation of the Magnetical Needle", p. 577.

[28] MacPike (1932), p. 229.

[29] Halley's manuscripts of this lecture, entitled "Concerning the motion of light by Mr Halley" (Royal Society RBC 7.391), and a related lecture, "Some observations on the motion of the sun" (Royal Society RBC 7.364) read on 18 October 1693, are edited by Schaffer (1977), pp. 29–33.

[30] Schaffer (1977), p. 30.

Table 2.1 Streete's analysis of the lunar eclipses reported by Ptolemy and al-Battānī

Date of eclipse	Difference in longitude between observation and calculation
19 March 721 B.C.	−12′57″
8 March 720 B.C.	+11′9″
1 September 720 B.C.	−28′11″
21 April 621 B.C.	+13′53″
16 July 523 B.C.	−30′6″
19 November 502 B.C.	+6′17″
25 April 491 B.C.	−6′4″
22 December 383 B.C.	−12′28″
18 June 382 B.C.	−13′54″
12 December 382 B.C.	−0′23″
22 September 201 B.C.	+8′27″
19 March 200 B.C.	+11′35″
11 September 200 B.C.	−15′7″
30 April 174 B.C.	+0′41″
27 January 141 B.C.	+3′30″
5 April 125 A.D.	−10′58″
6 May 133 A.D.	+8′47″
20 October 134 A.D.	+2′11″
5 March 136 A.D.	−25′57″
23 July 883 A.D.	+0′14″
2 August 901 A.D.	+5′58″

These eclipses had previously been analysed by Streete to test the accuracy of the tables in his *Astronomia Carolina*, first published in 1661. On page 98 of the *Astronomia Carolina* Streete listed 21 lunar eclipses recorded in Ptolemy's *Almagest* and al-Battānī's *De Scientia Stellarum*.[31] Converting the time of midpoint of each eclipse into equinoctial time at London, Streete calculated the longitude of the sun and moon and deduced the difference in longitude between the moon and the point 180° along the ecliptic from the sun, obtaining the results given in Table 2.1. The differences Streete found are fairly evenly distributed between positive and negative (9 negative against 12 positive) and are generally of the order of 10′ of longitude, which corresponds to an error in time of about 20 minutes.

In order to convert the local times of the eclipses to London time, however, Streete had needed the difference in geographical longitude between the places of observation and London. On page 120 of his book, Streete lists the latitudes and longitudes of 80 cities; 18 of these are marked with an asterisk to indicate that the geographical coordinates have been "gathered by more certain Cœlestial observation".[32] In Table 2.2 I give Streete's longitudes for the cities from which the eclipses reported by Ptolemy and al-Battānī were observed together with the modern longitudes of those cities (also given are the longitudes from the second and third editions of Streete's *Astronomical Carolina*, which, as we shall see, differ from those in the first edition). It is clear that Streete has considerable errors in his

---

[31] Streete clearly saw the importance of ancient and medieval observations in refining and testing his astronomical theories. In an undated letter to Edward Bernard, Savilian Professor of Astronomy at Cambridge and a competent Arabist, Streete asked Bernard to send any ancient observations that he found in Arabic sources. See Mercier (1994), pp. 190–191.

[32] Streete, *Astronomia Carolina*, p. 119.

Table 2.2 Streete's longitudes east of London for the cities in which the eclipses reported by Ptolemy and al-Battānī were observed

City	Modern	Steete (1st ed.)	Streete (2nd/3rd ed.)
Alexandria	2h0'	2h11'	2h1'
Antioch	2h25'	3h31'	2h35'
Aracta (Al-Raqqa)	2h36'	3h15'	2h45'
Babylon	2h58'	3h1'	3h1'
Rhodes	1h53'	2h25'	1h54'

longitudes: Antioch is placed more than 1 hour too far to the east, and Aracta (al-Raqqa) and Rhodes are both more than 30' too far to the east. This is especially obvious when we consider the order of the cities going east. The correct order should be Rhodes, Alexandria, Antioch, Aracta, and then finally Babylon; from Streete, however, we get Alexandria, Rhodes, Babylon, Aracta, and Antioch. Even without good knowledge of the precise longitudes of these cities it would have been clear that this order was incorrect: Antioch is, after all, close to the Mediterranean coast and Aracta well inland. Streete's errors in the placement of these cities would have translated into systematic errors in his comparison of the historical eclipse records with his tables of more than 30' in celestial longitude for the observation in Antioch and about 20' for the observations from Aracta and Rhodes.

Despite the well-known problems in determining geographical longitudes—problems which on land at least could be solved through the simultaneous timing certain celestial events such as eclipses of the moon and the subsequent comparison of the observed local times, if only arrangements could be made to make the observations—the longitudes of these cities given by Streete are particularly bad. Riccioli, in his *Almagestum Novum*,[33] had placed Alexandria 8', Aracta 48', Babylon 58', and Antioch 1h3'4" east of Rhodes—values, with the exception of Antioch, much closer to the modern longitude differences and, crucially, in the correct order running east to west. However, Riccioli put Rhodes about 30' too far east from Uraniborg (his base meridian). Similarly Kepler, in the *Rudolphine Tables*,[34] put Alexandria 12', Antioch 37', Aracta 51', and Babylon 1h15' east of Rhodes, quite good values, but also overestimated by about 30' the difference in longitude between Rhodes and Uraniborg (he gave a fairly good figure for the longitude difference between Uraniborg and London). Ptolemy in his *Geography*,[35] had put Alexandria 7'20", Antioch 45'20", and Babylon 1h21'20" east of Rhodes (Aracta is not included by Ptolemy)—closer still to the modern longitude differences and all in the correct order. Despite the different longitude differences given in these different works, astute readers of Streete's book would still have been suspicious of his table of cities.[36] In the second and third editions of the *Astronomical Carolina*, published in 1710 and 1716, many of the latitudes and longitudes in the table of cities are corrected to values much closer to the modern values.[37] However, Streete's

---

[33] Riccioli, *Almagestum Novum*, I, p. 249.

[34] Kepler, *Tabulæ Rudolphinæ, tabb. f.* 33–36.

[35] Ptolemy, *Geography*, 4,5,9 (Alexandria), 5,2,34 (Rhodes), 5,15,16 (Antioch), and 5,20,6 (Babylon); Stückelberger and Graßhoff (2006), pp. 423, 499, 567, and 593.

[36] It should be remarked, however, that some other tables of cities are as bad as Streete's. Vincent Wing, Streete's biggest critic, gave a slightly better longitude for Alexandria (2h3' east of London) but a much worse longitude for Babylon (3h40' east of London) in his *Astronomia Instaurata*, p. 133 (Wing does not list Antioch, Aracta, or Rhodes in his table).

[37] Although Halley is credited in the preface to the 2nd and 3rd editions of the *Astronomical Carolina* with revising some of the tables, it is not certain that the values given in the table of cities are due to Halley. The values are not identical with those given in Halley's posthumous astronomical tables.

comparison of his tables with the eclipses reported by Ptolemy and al-Battānī was not altered to take into account the corrections to the longitudes of the cities.

In his unpublished paper Halley remarked that when considering the observations reported by al-Battānī "our Astronomers have been forced to remove these places much more easterly than they are, and particularly our Mr Street (a man whose skill and industry hardly allowed him superior in this art) has been forced to commit a very great Absurdity in his Caroline Tables"[38] in altering the longitudes of Antioch and Aracta. If the cities are restored to their proper positions, then we must instead assume that the apparent motion of the sun is retarded over time. Finally, he refers to the moon's motion:

> Ptolemy makes Babylon too near Alexandria by 3/8 of an hour therefore in reducing the Babylonish observations to Alexandria he makes all of their times later than they were, and the interval between them and the observations made at Alexandria too little so that he makes the ☾ revolve in less time than (it) really did.

By the following year Halley had undertaken a further, more careful, analysis of the ancient observations reported by Ptolemy and those from al-Battānī. He had found that instead of showing a lengthening of the solar year, which would indicate that the Earth's motion was being retarded, the year was in fact getting *shorter*. Halley presented his results to the Royal Society on 18 October 1683, along with an explanation for why the Earth was accelerating not decelerating.[39] He argued that the orbit of the Earth will grow less and less over time, approaching the sun in a spiral, and so the year will grow shorter and shorter. The same argument can be applied to the moon, and Halley remarks that "whereas the moon in her motion about the earth seems not to have accelerated proportionately so much as the Sun is chiefly to be attributed to the slowness of the moons motion about the earth".[40]

Surprisingly, Halley does not mention the issue of the acceleration of the Earth or the moon in his paper containing his corrections to and study of *De Scientia Stellarum*, although this work provided the basis for his investigation of the phenomenon. In fact, Halley's only published comment concerning the secular acceleration appears at the end of a report on the ruins of Palmyra published in the *Philosophical Transactions* in 1695.[41] Noting that the present day city of Racca (al-Raqqa) on the Euphrates river is the city of Aracta where al-Battānī made his observations, Halley writes:

> The Latitude thereof was observed by that *Albatâni* with great accuranteness, about eight hundred years since; and therefore I recommend it to all that are curious of such Matters, to endeavour to get some good Observation made at this Place, to determin the Height of the *Pole* there, thereby to decide the Controversie, whether there hath really been any Change in the *Axis* of the Earth, in so long an Interval; which some great Authors, of late, have been willing to suppose. And if any curious Traveller, or Merchant residing there, would please to observe, with due care, the *Phases* of the *Moons Eclipses* at *Bagdat, Aleppo* and *Alexandria*, thereby to determine their Longitudes, they could not do the Science of *Astronomy* a greater Service: For in and near these Places were made all the Observations whereby the Middle Motions of the Sun and Moon are limited: And I could then pronounce in what Proportion the *Moon's* Motion does Accelerate; which that it does, I think I can demonstrate, and shall (God willing) one day, make it appear to the Publick.

Halley's promised demonstration never appeared. In part this may have been simply a lack of time. In August 1695, Halley offered to recalculate the elements of cometary orbits for inclusion in the second edition of Newton's *Principia*. This was a difficult and lengthy task, but it allowed Halley to take the first steps towards identifying the periodic return of what is now known as Halley's comet. In early 1696, Newton appointed Halley to be his deputy at the Chester Mint—a position that took up much of Halley's time for the next 2 years—following which Halley embarked on several long sea

---

[38] Schaffer (1977), p. 31.

[39] Schaffer (1977).

[40] Schaffer (1977), p. 32.

[41] Halley, "Some Account of the State of Palmyra".

Table 2.3 Halley's secular equations for Jupiter and Saturn

Year (A.D.)	Secular equation	
	Jupiter	Saturn
100	0°0.6′	0°1.4′
200	0°2.3′	0°5.6′
300	0°5.2′	0°12.5′
400	0°9.2′	0°22.2′
500	0°14.3′	0°34.8′
600	0°20.6′	0°50.0′
700	0°28.1′	1°8.1′
800	0°36.7′	1°29.0′
900	0°46.4′	1°51.6′
1000	0°57.3′	2°19.0′
1100	1°9.4′	2°48.2′
1200	1°22.6′	3°20.2′
1300	1°36.9′	3°55.0′
1400	1°52.4′	4°32.5′
1500	2°9.0′	5°12.8′
1600	2°26.8′	5°55.9′
1700	2°45.7′	6°41.7′
1800	3°5.8′	7°30.4′
1900	3°27.0′	8°21.8′
2000	3°49.4′	9°16.1′
2100	4°12.9′	10°13.1′
2200	4°37.5′	11°12.9′
2300	5°3.3′	12°15.4′
2400	5°30.3′	13°20.8′
2500	5°58.4′	14°28.9′
2600	6°27.7′	15°39.7′
2700	6°58.0′	16°53.4′
2800	7°29.6′	18°9.8′
2900	8°2.0′	19°29.1′
3000	8°36.1′	20°51.1′

journeys. In 1704 Halley was appointed Savilian Professor of Geometry in Oxford and immediately returned to work on the lunar theory. However, no references to the secular acceleration appear in any of his later publications or in known manuscript sources. Halley had not forgotten the issue of long-term changes in the mean motion of bodies in the solar system, however. By about 1720 he had compiled a new set of astronomical tables. For reasons that are not fully known, these tables were only published posthumously by John Bevis in 1749 and again in 1752 with an English translation of the precepts.[42] Neither the solar nor the lunar tables make any allowance for a secular acceleration, but Halley does provide tables to take into account the secular accelerations of Jupiter and Saturn. Both tables cover the period from A.D. 100 to A.D. 3000 and correctly give the correction to the mean motion of the planets—referred to as the "secular equation" (Æquatio Secularis)—using a quadratic function (see Table 2.3). For Jupiter the correction is added to the mean motion and for Saturn it is subtracted, implying that one planet is accelerating but the other decelerating. Halley gives no explanation for how he has determined these secular equations. Their presence in his tables, however,

---

[42]Cook (1998), pp. 354–376.

raises the question of why he does not provide similar secular equations for the sun and moon, having claimed the discovery of both. I suspect the answer is that Halley was unable to satisfactorily determine the magnitude of the lunar and solar accelerations. The available solar data from antiquity consisted almost wholly of the equinoxes and solstices found in Ptolemy's *Almagest*, which were known to be problematical, and his analysis of the ancient and medieval eclipses was hampered by lack of accurate determinations of the geographical longitudes of the places of observation. An essential part of Halley's argument when he had announced his discovery of the moon's acceleration was that the longitudes given by Streete and others were in error. Although his own longitudes for these cities might well have been better, they were still somewhat uncertain and could easily have been challenged by other astronomers. Without the means to solve these problems Halley may have been unwilling to commit himself to a value for the size of the accelerations which he knew he could not adequately defend.

Two later authors claimed that Halley had undertaken further work on the secular acceleration. In 1736 William Whiston claimed to report a determination of the magnitude of the secular acceleration by Halley, but it may be that this was Whiston's own estimate rather than Halley's (see Chap. 3). In 1757 Jérôme Lalande said of Halley's work that in order "to make this research useful, it was necessary to have the longitudes of the places where Albategnius observed; I think that after the observations that M. de Chazelles made in Alexandrette in 1694, M. Halley concluded more positively this acceleration, at least if we are to believe a passage in the second edition of M. Newton".[43] Newton's words, however, refer only to Halley's discovery of the secular acceleration and imply nothing further than Halley had announced in his 1695 paper (see below).

Early Reaction to Halley's Discovery of the Moon's Secular Acceleration (1695–1734)

Halley's announcement of the discovery of the secular acceleration of the moon elicited almost no response from the English astronomical community. Despite the importance of lunar theory, the discovery of a new variation in the moon's motion was met with widespread silence. Nothing published in the *Philosophical Transactions* over the next fifty years referred to Halley's discovery and many important astronomical texts such as Charles Brent's *The Compendius Astronomer*, John Keill's *An Introduction to the True Astronomy*, Robert Wright's *New and Correct Tables of the Lunar Motions*, or the series of books of Charles Leadbetter (*A Complete System of Astronomy, Uranoscopia, Astronomy*, etc.) made no mention of it.

The first allusion to Halley's discovery appears to have been made by David Gregory in his *Astronomiæ Physicæ & Geometricæ Elementa* (1702; English translation published as *The Elements of Astronomy, Physical and Geometrical* in 1715). Gregory had been preferred to Halley in the election for the Savilian chair of astronomy at Oxford in 1691, but the two remained respectful of each other's work, and when Halley was elected to the Savilian chair of geometry the two collaborated on an edition of Apollonius, and Halley contributed a "Synopsis of the Astronomy of Comets" to the English edition of Gregory's *Elements of Astronomy*.[44] In book IV, proposition 24 of the *Elements of Astronomy*, after listing the various anomalies in the motion of the moon, Gregory remarked:

> And at length by comparing the Period of the Moon which is now observ'd, with that that was known several Ages ago by their observations, whether the Globe of the Earth is in this process of Time encreas'd or dimishish'd,

---

[43]Lalande, "Mémoire sur les Équations Séculaires", p. 426.
[44]Cook (1998), p. 323.

(as I suspect) and Geometrically determine, by *Schol. Prep.* 17. what Increment or Decrement has happen'd in the whole time.[45]

Gregory's argument here is that if the size of the Earth changes over time then this will affect its gravitational pull on the moon, causing the moon's motion to accelerate or decelerate. He continued by speculating that there may be yet further inequalities in the moon's motion that have not yet been discovered:

> But if other Inequalities, hitherto not observ'd, affect the Moon's Motion, taking away the foremention'd ones, they will appear, and being made plain will encrease our Astronomy, by making up the Moon's Theory.[46]

The first explicit reference to Halley's discovery of the secular acceleration of the moon appears in the second edition of Newton's *Principia*. Proposition 42 of book 3 ends with a discussion of comets approaching closer to the sun on every passage, because the sun possesses an atmosphere which causes resistance. Eventually, the comet will fall into the body of the sun. Newton then considers the possibility of comets around the fixed stars, renewing that star and causing it to shine brighter, which cause he attributes to the sightings of the new stars of 1572 and 1604. In the second edition of 1713, Newton continues with the remark (similar to Gregory's) that as the body of the sun reduces in size, the mean motions of the planets will decrease, and as the size of the earth increases, the mean motion of the Moon will increase

> And by combining the eclipse observations of the *Babylonians* with *Albategeni*'s and with today's, our *Halley* showed that the mean motion of the Moon when compared with the diurnal motion of the Earth, gradually accelerates, as far as I know the first of all to have discovered it.[47]

This remark is deleted in the third edition for reasons that are not understood. In 1734 William Whiston noted his puzzlement on their omission in the later edition: "Which Words yet are drop'd in the Third Edition of that Book: By what Means I cannot certainly tell".[48] In 1757 Jérôme Lalande suggested a possible motivation for Newton's removal of this sentence in a change in Newton's thinking on the issue of comets:

> But the passage was completely eliminated in the third edition, and one cannot tell whether it is because M. Newton was suspicious of Halley's opinion, or if it is because he himself challenged the explanation that he seemed to want to give. This explanation consisted of assuming that the vapours of the sun and comets join with our atmosphere, increasing the mass of the Earth and so making the central force greater, the orbit contracts and has a lesser period of revolution.[49]

Another possible explanation is simply that Newton felt that he was getting away from the main content of the proposition by bringing in the issue of the motions of the heavenly bodies. He may have considered that this was particularly an issue as the passage appeared just before the General Scholium which ends the *Principia*. Newton evidently considered that the General Scholium, in which he discusses God and hypotheses, was an important part of the work, for he worked through several drafts in composing the final version, and he was very careful in his choice of words.[50]

---

[45] Gregory, *The Elements of Astronomy*, p. 540.

[46] Gregory, *The Elements of Astronomy*, p. 540. See also his comments on p. 552: "But also when other Errors are discover'd, other Tables must be constructed, fully and perfectly to rectify the Moon's Motion, which is far the hardest Work in Astronomy. Let what we have said of the Order of these Tables, suffice in the mean time".

[47] Newton, *Principia*, 2nd ed., p. 481.

[48] Whiston, *Six Dissertations*, p. 158.

[49] Lalande, "Mémoire sur les Équations Séculaires," p. 426.

[50] Cohen and Whitman (1999), pp. 274–292.

Chapter 3
A Forgotten Episode in the History of the Secular Acceleration: William Whiston, Arthur Ashley Sykes and the Eclipse of Phlegon

The Eclipse mentioned by Phlegon, in the thirteenth book of his Olympiads, and applied by all the most learned defenders of Christianity to the Darkness which happen'd at the Passion of our Saviour, having always appeared to me to be one of the most unexceptional external proofs that we have handed down to us for any of the facts recorded by the Evangelists, and not meeting with any reasons in your Dissertation, and Defence of it on this subject, to alter my opinion, I think proper to offer you some remarks upon your performances ... without medling with your rights and reasons for entering into this controversy; or any of those accidental differences betwixt you and Mr. Wh—n, which might afford some entertainment to common readers, but are entirely foreign to the matter in debate.

—Anonymous, *Phlegon's Testimony Shewn to Relate to the Darkness Which happened at our Saviour's Passion, In a Letter to Dr. Sykes* (London, 1733).

In 1734 William Whiston published *Six Dissertations* dealing with biblical history and astronomical chronology. Tucked away in a long discussion of the calculation of ancient eclipses, Whiston referred to Halley's discovery of the moon's secular acceleration, remarking that "This I have all along esteemed one of Dr. *Halley*'s greatest Discoveries in *Astronomy*",[1] and provided an estimate of the correction to be applied to the calculation of the time of syzygy on account of the moon's secular acceleration, the first correction for the secular acceleration to appear in print. Whiston's discussion of the secular acceleration was presented within an ongoing acrimonious dispute between Whiston and Arthur Ashley Sykes, a Cambridge educated clergyman and prolific writer of controversial religious pamphlets, over whether the eclipse reported by Phlegon should be associated with the darkness during Christ's crucifixion, and appears to have passed other astronomers by without making any impression. As we will see, only one other scholar, George Costard made any reference to Whiston's discussion of the secular acceleration, which quickly became a forgotten episode in the history of its study, missing from all later (and contemporary) accounts of the subject.

---

[1] Whiston, *Six Dissertations*, p. 157.

J.M. Steele, *Ancient Astronomical Observations and the Study of the Moon's Motion (1691–1757)*, Sources and Studies in the History of Mathematics and Physical Sciences, DOI 10.1007/978-1-4614-2149-8_3, © Springer Science+Business Media, LLC 2012

Dramatis Personæ

William Whiston and Arthur Ashley Sykes shared a propensity to become involved in heated and often controversial debates on scripture and the church. Whiston, 17 years the senior, was born into a clerical family in Norton-juxta-Twycross in Leicestershire on 9 December 1667.[2] He was educated at Clare Hall, Cambridge receiving his BA in 1690 and then MA in 1693 and was elected to a Probationary Fellowship at Clare the same year. Also in 1693 Whiston was ordained and the following year was appointed Chaplin to Bishop John Moore. Both as a student and Fellow at Clare, Whiston displayed an interest in mathematics and the new philosophy of Newtonian physics. Drawing inspiration from Newton's work on comets, Whiston's *A New Theory of the Earth*, published in 1694, applied Newtonian physics and astronomy to explain God's creation (and eventual destruction) of the solar system. In 1701, Whiston lectured at Cambridge as Newton's deputy, and following Newton's resignation from the Lucasian professorship, Whiston was elected to the chair in 1702. Whiston gave lecture courses on astronomy and Newton's mathematical philosophy, both of which were later published in Latin and English versions.

During his years at Cambridge Whiston formed friendships with Samuel Clarke and other latitudinarian divines. Whiston published several works on questions of religion, focusing particularly on the question of prophecy and the revival of "Primitive Christianity" or the doctrines of the early Church. Whiston's uncompromising views led in 1710 to his being expelled from Cambridge on account of his antitrinitarian preaching, following a bitter trial about which Whiston felt betrayed by his friends and the university. Whiston had to turn to public teaching in London to earn his living. In London Whiston set up the Society for Primitive Christianity and continued to publish voluminous works on the subject, including many studies and translations of relevant ancient authors (most significantly his translation of Josephus, which remained the standard translation until the late 1800s and is still in print today). He remained an active, if at times shunned, member of the scientific community in London, regularly attending Royal Society meetings. Among his scientific work, the most significant lay in raising the issue of the determination of longitude at sea. In 1714 Whiston and Humprey Ditton successfully petitioned Parliament to set up a prize for solving the problem of determining the longitude. Whiston proposed several solutions to the problem, some quite outlandish, but never succeeded in being awarded the prize. Whiston continued his scientific work until late in his life, never giving up on his hope of returning to an academic career and even considering applying for the position of Astronomer Royal on the death of Flamsteed until he heard that Halley had already been nominated.

Whiston became embroiled in several acrimonious disputes during his time in London. He launched an attack on Isaac Newton's chronology (although judiciously waiting until after Newton's death to publish), fought Henry Sacheverell who had tried to have Whiston excluded from St Andrew's Church in Holborn, claimed that an account of Archbishop Cranmer's recantation was not genuine, and had several battles with Arthur Ashley Sykes and, most notably, Anthony Collins. Whiston was firm in his arguments and uncompromising in the face of opposing views. Without a recognised position, these arguments also served to keep Whiston in the public eye, no doubt helping attract students for his lectures and ensuring an audience for his books and pamphlets, many of which were published by his son John, and which provided part of Whiston's livelihood. Whiston's argumentative style, religious views, and apocalyptic prophecies were targeted by the Scriblerians, an informal group of wits including

---

[2] Detailed biographies of Whiston may be found in Farrell (1981) and Force (1985), pp. 10–31. Both these works make extensive, and generally uncritical, use of Whiston's autobiographical *Memoirs of the Life and Writings of Mr. William Whiston Containing Memoirs of Several of his Friends Also*, published in 1749, 3 years before his death. Buchwald and Feingold (forthcoming), Chap. 8, provides a more reasoned account based upon a wider variety of source material.

Pope, Swift and Gay,[3] culminating in 1731 in the publication of *Whistoneutes: or, Remarks on Mr. Whiston's Historical Memoirs of the Life of Dr. Samuel Clarke, &c*, an anonymous satirical work published under the pseudonym Simon Scriblerus which ridiculed Whiston.[4] While the attacks may have been painful for Whiston, they probably also provided him with some notoriety, which again may have helped sales of his own works.

Like Whiston, Arthur Ashley Sykes was born into a clerical family.[5] Sykes entered Coprus Christi College, Cambridge, in 1701, graduating with a BA in 1705 and MA in 1708, and receiving a DD in 1726. In 1713 Sykes was given the vicarage of Godmersham, Kent, was presented to the rectory of Dr. Drayton the following year, and in 1718 to the rectory of Rayleigh, Essex, where he remained until his death in 1756. It is not known whether Sykes attended any of Whiston's lectures at Cambridge, but both were friends of Samuel Clarke and certainly knew one another. Like Whiston, Sykes was a prolific author who was not afraid of controversy. His early pamphlets included energetic defences of baptism by dissenters, a defence of Samuel Clarke's Arianism, and arguments in support of latitudinarianism within the Church of England. Unlike Whiston, however, Sykes managed to avoid open confrontation with those who did not share his religious views. Whilst he could present his views just as forcibly as Whiston, Sykes's writing could also be humorous and was often less hectoring than Whiston's.

Through Samuel Clarke, Sykes was appointed afternoon preacher at King Street Chapel, Golden Square and subsequently morning preacher and assistant preacher at St James's, Westminster, alongside his rectory at Rayleigh. This allowed Sykes to maintain personal contact with many Cambridge and London intellectuals, most notably Isaac Newton. Following Newton's death, Sykes was invited to edit some of Newton's theological works for publication, but nothing came of this project.

Whiston, Sykes, and the Eclipse of Phlegon

In 1705 Samuel Clarke gave the tenth Boyle lectures at the Cathedral Church of St Paul. These annual lectures were endowed through the estate of the English natural philosopher Robert Boyle with the intention of providing a forum for the Anglican church to defend Christianity.[6] Clarke, who had also given the lectures the previous year, took as his topic for 1705 the evidence for natural religion and the Christian revelation.[7] Over the course of eight sermons Clarke discussed the relationship between God and rational creatures, man's moral obligations, the insufficiency of Greek and other heathen philosophy for the reforming of mankind, man's need for divine revelation, and the truth of the Bible. Discussing the account of the life and death of Jesus given in the gospels, on page 325 of the published version of the lectures Clarke mentioned that the earthquake and the miraculous darkness that took

---

[3] Rousseau (1987).

[4] This work was probably not written by one of the original Scriblerians (Pope, Swift, Gay, Arbuthnot, and Parnell); the "Scriblerus" appellation seems to have been adopted by many satirical writers of the eighteenth century. See Marshall (2008).

[5] Biographical information on Sykes may be found in Disney, *Memoirs of the life and writings of Arthus Ashley Sykes*, on which the following account is based.

[6] On the early history of the Boyle lectures, see Dahm (1970).

[7] Clarke's lectures were published with the title *A Discourse Concerning the Unchangeable Obligations of Natural Religion, and the Truth and Certainty of the Christian Revelation* in 1706 and then republished together with his 1704 lectures under the title *A Discourse Concerning the Being and Attributes of God, the Obligations of Natural Religion, and the Truth and Certainty of the Christian Revelation*, which was reprinted at least ten times during the eighteenth century. On Clarke's lectures, see Dahm (1970) and Gascoigne (1989), pp. 117–126.

place during the Passion are also described in histories written by Roman authors, providing an independent confirmation of the biblical account:

> The Crucifixion of Christ under *Pontius Pilate*, is related by *Tacitus*. And diverse of the most remarkable Circumstances attending it, such as the *Earthquake* and miraculous *Darkness*, were recorded in the publick Roman Registers, commonly appealed to by the first Christian Writers as what could not be denied by the adversaries themselves; and are in a very particular manner attested by *Phlegon*.[8]

Clarke accompanied this passage with a quotation from a fragment of Phlegon referring to an eclipse of the sun in the fourth year of the 202nd Olympiad. Early Christian authors such as Julius Africanus and Eusebius associated this eclipse with the darkness at the Passion. The account of Phlegon's eclipse was not without problems, however. First, since the Passion took place at the Jewish Passover, this should have been on or around the full moon, in which case a solar eclipse would not be possible. However, if the eclipse was a supernatural event then the laws of astronomy need not hold (this, as we shall see, was Whiston's view). The early Christian historian George Syncellus inserted into Phlegon's account a remark that the eclipse took place at full moon, confirming its miraculous status, but as Clarke and others knew that these words were missing from the other ancient accounts. Secondly, the preserved sources for Phlegon's eclipse contained discrepancies in the year in which the eclipse was supposed to take place: Philoponus gave the second year of the 202nd Olympiad whereas Eusebius gave the fourth year of the 202nd Olympiad. Furthermore, calculation of the solar eclipses that would have been visible to during the 202nd Olympiad showed that the eclipse must have taken place during the first year of that Olympiad. Thus scholars hoping to use this eclipse to establish the date of the death of Christ were faced with having to either correct the preserved fragments or concluding that the Christian chroniclers, upon whose writings much of early A.D. chronology relied, did not have full command of ancient calendars and eras. Unwilling to admit that Phlegon and other ancient astronomers were not as good as he took them to be, the sixteenth century chronologer Joseph Scaliger simply omitted Phlegon's eclipse from his discussion of the Olympiads in his *Anagraphe*.[9] Kepler, on the other hand, argued in favour of correcting the text of Phlegon to eliminate the problem. In the *Eclogæ Chronicæ* of 1615 Kepler suggested emending (some would say contorting) the text to take the numeral letter δ as a mistake for the ending δὲ, and to translate the passage as "in the year of the 202nd Olympiad", indicating the first year of the 4-year cycle.[10] Kepler noted in the margin that he has calculated that a solar eclipse took place at about noon or 2 hours before on 25 November of the corresponding year (A.D. 29) and that the eclipse was almost total in Asia.[11] These issues were probably of no concern to Clarke, however: for him the association of the eclipse of Phlegon with the darkness at the death of Jesus was well known and uncontroversial. The account of the eclipse provided a useful confirmation of the biblical account and no more needed to be said about it.

Clarke's Boyle lectures were reprinted several times during his lifetime. Shortly before his death, Arthur Ashley Sykes told Clarke that he believed that the common interpretation of the eclipse of Phlegon as being an account of the darkness at the crucifixion was unjustified. Sykes had a long-standing friendship with Clarke and had benefited from Clarke's patronage when he was appointed preacher at King Street Chapel and then St James's in Westminster, and in a number of publications Sykes had defended Clarke's views on the Trinity. Sykes told Clarke that the eclipse of Phlegon was

---

[8] Clarke, *A Discourse Concerning the Unchangeable Obligations of Natural Religion, and the Truth and Certainty of the Christian Revelation*, p. 325.

[9] Grafton (1993), p. 559.

[10] Kepler, *Eclogæ Chronicæ*, pp. 87, 126.

[11] Kepler, *Eclogæ Chronicæ*, p. 87. Kepler's dating of the eclipse formed the basis of almost all further work on this eclipse until today. For recent studies of this eclipse, see Fotheringham (1920), Newton (1970), pp. 110–113, Stephenson (1997), pp. 359–360, and Smith (2008).

in his view probably a real eclipse (as had also been assumed by Kepler when he dated it), and that it was simply coincidence that it happened around the time of the death of Jesus. As such it could not be used as a confirmation of the accuracy of the biblical account. Sykes succeeded in convincing Clarke that the discussion of Phlegon's eclipse should be removed from future editions of his lectures. The matter would almost certainly have rested there had it not been for William Whiston. Following Clarke's death, Sykes published a memoir of Clarke's life. Whiston, also a friend of Clarke, also decided to publish a memoir to supplement Sykes account (and also that given by Bishop Hoadley in another eulogy) since "I, who knew Dr. *Clarke*, his Character, Writings, and Conduct, long before Dr. *Sykes*, and much more authentickly, in many Points, than either Dr. *Sykes*, or Bishop *Hoadley*, and in some Points better than his own Brother, Dean *Clarke*".[12] In his *Memoir*—a work that is as much about Whiston himself as it is about Clarke—Whiston recounted the story of Sykes convincing Clarke to remove the passage about Phlegon from future editions of his Boyle lectures:

> Some time before Dr. *Clarke* died, Dr. *Sykes* persuaded him to leave out of the future Editions of his *Boyle*'s *Lectures*, that famous Passage in *Phlegon* of an Eclipse of the Sun, and an Earthquake, which was cited by him, and has been generally cited others of the Learned, as an Attestation to the supernatural Eclipse of the Sun, and the Earthquake at our Saviour's Passion, mentioned by the Evangelists. When I came to enquire of Dr. *Sykes* his Reasons for such his Persuasion of Dr. *Clarke*, I found it was only a *Supposal*, that some natural Solar Eclipse or other might be fitted to some Earthquake in *Bithynia*.[13]

Whiston continued by explaining that he had calculated the circumstances of all eclipses that could have happened during in any of the years Phlegon could mean and found that "no *Natural* Eclipse of the Sun could possibly happen so as to suit his Description, but only that *Supernatural* one at the Passion".[14] Whiston had long argued that the Bible contained evidence not only of miracles but also of the fulfillment of prophecies. In 1707 Whiston gave the Boyle lectures on the subject of the fulfillment of biblical prophecy, both in ancient and more recent times. For Whiston the eclipse of Phlegon was important not only in supporting the biblical account of the crucifixion but also in confirming Old Testament prophecy. Jeremiah had foretold a solar eclipse at the death of Christ, and both the gospels and Phlegon confirmed that the eclipse had happened. In Whiston's opinion the eclipse was a miracle, evidence that God was not restricted by the laws of nature. As such it was important that the eclipse was *not* a natural event, and therefore Phlegon's eclipse must not be a natural eclipse that could be calculated using astronomical theory. Thus Whiston's analysis of calculated ancient eclipses was in order to prove that Phlegon's eclipse could not be matched with one of these eclipses.

Whatever Whiston's intentions concerning the eclipse, Sykes was outraged by Whiston's account, describing its author as a "man of warmth, and zeal, and indiscretion, (who) has taken a matchless liberty to misrepresent and abuse myself as well as others; to pass hard censures upon some facts which he knew not the true grounds of; and to relate some direct falsehoods, without consulting, or regarding, the only persons that could have set him right".[15] Sykes responded with a *Dissertation on the Eclipse Mentioned by Phlegon* in which he argued that the eclipse of Phlegon was a natural eclipse. After recounting the ancient accounts of the eclipse, he discussed Kepler's claim that eclipse should be dated to 24 November 29 A.D. Sykes said that he has asked Whiston to calculate the circumstances of this eclipse, and published Whiston's result. He quoted Whiston as saying that the

---

[12] Whiston, *Memoirs of the Life of Dr. Samuel Clarke*, p. 4.

[13] Whiston, *Memoirs of the Life of Dr. Samuel Clarke*, p. 148.

[14] Whiston computed the circumstances of the eclipses using his "Copernicus", a mechanical instrument he had devised for calculating astronomical phenomena. The Copernicus instruments were made by John Senex and sold both by Senex and by Whiston himself at the price of six guineas. Whiston presented the instrument at a meeting of the Royal Society on 10 February 1715/6. According to Farrell (1981), p. 216 no examples of Whiston's "Copernicus" have been identified. Whiston also wrote *The Copernicus Explained* (1715), describing the instrument and its use.

[15] Sykes, *Dissertation on the Eclipse Mentioned by Phlegon*, p. 4.

eclipse does not agree with the description given by Phlegon, but said that the only objection that he himself can see is the minor one that Whiston's calculated eclipse gives the time of the eclipse somewhat earlier in the day than is reported by Phlegon.[16] Thus, in Sykes's view, Phlegon is describing a natural eclipse that took place some years before the crucifixion.

The same year as Sykes's *Dissertation*, and almost certainly in response to it, although he does not mention Sykes's work, Whiston published *The Testimony of Phlegon Vindicated: or, An Account of the Great Darkness and Earthquake at Our Savior's Passion, Described by Phlegon*. In it, he argued against Kepler's identification of the eclipse, noting that Kepler had been forced to alter Phlegon's account that the eclipse was in the fourth year of the 22nd Olympiad, and further that the time of day at which Kepler's eclipse of A.D. 29 took place was off by three of four hours. In a scarcely veiled attack on Sykes, Whiston wrote:

> This being the true and certain state of the facts and testimonies, as to the eclipse mentioned by *Phlegon*, in the first six centuries; and the pretences against its application to the darkness at Christ's passion, being so intirely groundless, though lately said to be *almost certain*, it remains to me a very difficult problem, how that great and judicious person, Dr. *Clarke*, should so easily be persuaded to give up *Phlegon*'s testimony, upon the producing of *Kepler*'s hypothesis; supposing it was produced to him. … In this vehemence of *Kepler*'s *inclinations*, I do not so much wonder at his determinations. We see every day what mighty things *strong inclinations* can do. But knowing Dr. *Clarke* had no violent passions at all, and not believing he had any particular *inclination* to set aside *Phlegon*'s testimony: especially not to leave it out of his book in the *eighth*, after it had stood there *seven* editions: I cannon solve this problem; How *Kepler*'s eclipse, so imperfectly stated by him, and of so little consequence when stated to the best advantage, could prevail upon Dr. *Clarke* to discard it. … So that I must intirely leave this problem to such as can better account for it; it being still, I confess, to me utterly insoluble.[17]

Although Sykes was not mentioned by name, readers of Whiston's text would have known that it was he whom Whiston was referring to as having persuaded Clarke to change his view with respect to the eclipse. Whiston had named Sykes as this person in his earlier *Memoir of the Life of Dr. Samuel Clarke*, and this would not have been lost on his audience. Nor was it lost on Sykes. The following year he published *A Defence of the Dissertation on the Eclipse Mentioned by Phlegon: Wherein it is further shewn, That the Eclipse had no Relation to the Darkness which happened at our Saviour's Passion: And Mr. Whiston's Observations are particularly considered*, in which he complained that:

> Mr. *Whiston* has since published a Pamphlet under this title, *The Testimony of* Phlegon *vindicated; or, an account of the great darkness and earthquake at our Saviour's passion described by* Phlegon: *including all the testimonies of both* Heathen *and* Christian *in the very words of the original authors, during the first six centuries of Christianity, with proper observations on those testimonies*. The public *advertisments* added, (I suppose by Mr. *Whiston*'s order) that "all Dr. *Sykes*'s arguments to the contrary are fully confuted" in it. When I came to peruse this treatise, and found my name neither mentioned in the title page, nor once in the book itself; when I found not one argument which I had urged in the Dissertation confuted; no comparing together what different authors had said, in order to fix the *year* in which *Phlegon*'s eclipse happened; no enquiry into the probabilities of the account; no instance produced of a similar manner of expression in any Heathen writer whatever, where an *extraordinary darkness* was called an *eclipse of the Sun*; no account of the universal silence in *Greek* and *Latin* writers of that time about so remarkable an event; in short, no one thing done to remove or explain any one difficulty proposed; I could not but stand amazed at the publication of such an *advertisment*.[18]

In the *Defence*, Sykes reiterates his argument that the eclipse reported by Phlegon was not a supernatural eclipse and did not take place on the day of the crucifixion. Sykes says that he would have let the matter rest "had not a Gentleman, (to whom I was a perfect stranger) obliged me with a much more exact and true calculation of the eclipse in debate, than that which I published from Mr. *Whiston*'s authority".[19] Whiston's calculation had placed the eclipse at nine o'clock in the morning, whereas the

---

[16] Sykes, *Dissertation on the Eclipse Mentioned by Phlegon*, pp. 75–76.

[17] Whiston, *The Testimony of Phlegon Vindicated*, pp. 49–50.

[18] Sykes, *A Defence of the Dissertation on the Eclipse Mentioned by Phlegon*, pp. 1–3.

[19] Sykes, *A Defence of the Dissertation on the Eclipse Mentioned by Phlegon*, p. 3.

account of Phlegon said that the eclipse took place at the sixth hour of day, that is, at noon. Damning with faint praise, Sykes says:

> This was a material difference; and I took it for granted, that so *reputed* an astronomer as Mr. *Whiston*, could not have easily mistaken *three* or *four* hours in a computation of this kind. I imagined therefore the numbers in *Phlegon* rather to have been false, than Mr. *Whiston* to err; and I concluded that we "ought not to reject a certainty for so little an inaccuracy, as three or four hours, when all other circumstances agreed so exactly."[20]

However, the unknown gentleman, who it turns out had also written to Whiston signing his letter "G. M.",[21] had calculated that the eclipse reached totality just as Phlegon had described it around noon. Sykes reports that another gentleman, "whose skill in astronomical computations no one can question",[22] and who Sykes later reveals to be Mr. Hodgson, mathematics master of Christ's Hospital,[23] had also sent him calculations of the eclipse which agreed almost exactly with the previous gentleman's. Finally, Sykes's says,

> I consulted a *third* person, whose authority alone will instantly silence all doubts in a thing of this nature, and whose consummate knowledge in geometry and astronomy the whole world acknowledges, the great Dr. *Halley*.[24]

All three sets of calculations agreed that the eclipse took place a little after noon, in good agreement with Phlegon's account, but 3 hours later than Whiston had calculated. In calling on that authority of Halley, Sykes challenged not only Whiston's historical interpretations but also his astronomical competence. Whiston's reputation as an astronomer had suffered since his days as Lucasian professor at Cambridge. A series of ever more ingenious but impractical proposals for solving the longitude problem,[25] ridicule at the hands of "Simon Scriblerus",[26] and Whiston's need to earn a living through his astronomical lectures and his popular publications meant that he was never accorded the same respect as Halley or Flamsteed. Whiston had no choice but to respond to Sykes's attack.

Faced with the accusation that his own calculations were inaccurate, and still needing to argue that the eclipse of A.D. 29 could not have been the eclipse reported by Phlegon, Whiston now invoked the secular acceleration of the moon. Sykes had used Halley's authority to criticize Whiston, and Whiston decided he would cite Halley's own discovery to criticize Halley. Whiston had not previously mentioned the moon's acceleration in any of his published works: his lectures on astronomy and on Newton's *Principia* do not mention the phenomena, his "Copernicus" instrument did not make allowance for it, and the tables of the moon's mean longitude at the end of his *Copernicus Explained* are calculated without any correction for the acceleration. In the third of his *Six Dissertations*, entitled "A reply to Dr. SYKES'S Defence of his Dissertation on the Eclipse mentioned by Phlegon", Whiston began by explaining that the "Copernicus" allows only the approximate time of an ancient eclipse to be calculated:

> Since I published my *Vindication of the Testimony of* Phlegon, as relating to the *supernatural* Darkness and Earthquake at our Saviour's Passion, the Publick has been made acquainted with three distinct accurate

---

[20] Sykes, *A Defence of the Dissertation on the Eclipse Mentioned by Phlegon*, p. 4.

[21] Whiston, *Six Dissertations*, p. 134. BL MS Add 4224 ff. 218–222 preserves an account of the dispute over the eclipse of Phlegon written by Birch, based in part upon information provided directly by Sykes. F. 220 identifies G. M. as Gael Morris, assistant at the Greenwich Observatory.

[22] Sykes, *A Defence of the Dissertation on the Eclipse Mentioned by Phlegon*, p. 4.

[23] Sykes, *A Defence of the Dissertation on the Eclipse Mentioned by Phlegon*, p. 63.

[24] Sykes, *A Defence of the Dissertation on the Eclipse Mentioned by Phlegon*, p. 4.

[25] Farrell (1981), pp. 116–178.

[26] Farrell (1981), p. 36.

Calculations, from the best Astronomical Tables, or the Result of such Calculations, by a nameless Author, by Mr. *Hodgson*, and by Dr. *Halley*, of a *natural Eclipse*, supposed by *Kepler* to be that meant by *Phlegon*: As well as it had before been made acquainted, though without my Consent or Privity, of an inaccurate *Calculation*, or rather *Approximation* of mine, by an Instrument called by me the *Copernicus*. This I reckon does rarely err much about an Hour in the Time of the middle of the general Eclipse, and but proportionably in other Circumstances: As indeed it did scarcely err more than an Hour and twenty Minutes from Dr. *Halley*'s Calculation in the present Eclipse. This is to me a very ready, and a very useful Instrument upon many such Occasions: I mean where a greater Nicety is not necessary; which I took to be the Case as to that Eclipse of *Kepler*'s.[27]

Whiston's claim that the result of calculations by the Copernicus are approximate and to be used only in cases where precision is not necessary seems at odds with his statement in *The Copernicus Explained* that pointed to its application in solving chronological problems:

Since the Computation and Exhibition of all Eclipses, past or future, is by this Instrument become now so very easy, it will be fit to examin thereby all the old Eclipses men[t]ioned by Historians, and to compare them with Original Accounts, for the settling of Ancient Chronology and History; which Design was the very Occasion of the Contrivance of the same.[28]

Since the exact circumstances of the eclipse of A.D. 29 have now become important, Whiston says he will now give more accurate calculations of the eclipse. Whiston presents calculations of all of the possible solar eclipses between A.D. 29 and A.D. 33 to show that, in his view, none of the eclipses could have been that described by Phlegon. Whiston does not explain how he has calculated the eclipses, but it is possible to reconstruct his procedure from his results. The starting point for his calculations were the set of solar and lunar appended by Whiston to his *Astronomical Lectures*.[29] These tables were constructed by Flamsteed, who uncharitably wrote in a letter to Abraham Sharp that Whiston's lectures were "a very poore and barren piece: save that at the latter end he has added my new solar and lunar Tables".[30] Figure 3.1 shows Whiston's calculation of the eclipse of 24 November 29 A.D. which had been identified by Kepler as that described by Phlegon. Whiston begins in the top left by determining the mean elongation of the moon from the sun for the beginning of November 29 A.D. He calculates this elongation using the table of mean motion of the moon from the sun, given on page 42 of the appendix of his *Astronomical Lectures*, giving the result as $2^S4°46'16''$. Since he is interested in the conjunction later that month, Whiston's next step is to calculate the number of days, hours, minutes, and seconds until the next conjunction at which moment the moon's elongation from the sun will be $0^S0°0'0''$. Subtracting $2^S4°46'16''$ from $12^S0°0'0''$, Whiston obtains $9^S4°56'16''$. The closest value for the mean motion in a whole number of days found in the table on page 43 of his *Astronomical Lectures* is $9^S22°34'40''$ corresponding to 24 days. Subtracting $9^S22°34'40''$ from $9^S4°56'16''$ Whiston obtains $2°29'4''$. Using the same table he shows that this corresponds to 4 hours 53'30'', thus mean conjunction took place at 4 hours 53 minutes and 30 seconds after the noon preceding the 24 November 29 A.D. This time is given in the right-hand column of Whiston's calculation. Whiston now corrects this time by subtracting 31' for the effect of the moon's secular acceleration and adds 2h10'0'' for the difference in longitude between Greenwich and Nicaea to arrive at a local mean time of 6h32'30'' after the preceding noon. So far, Whiston has been working only with mean motion. In order to now obtain the time of true conjunction, Whiston calculates the true position of the sun and moon for the moment of mean conjunction using the tables on pages 6–34 of his *Astronomical Lectures*, and calculates the difference in longitude between the sun and the moon. The results of these calculations, which have been rounded to 5' or 10', are given in the top of the right-hand column,

[27]Whiston, *Six Dissertations*, p. 133.

[28]Whiston, *Copernicus Explained*, p. 43.

[29]Whiston's astronomical lectures were published in Latin as *Praelectiones Astronomicae Cantabrigaiae in Scholis Publicis Habitae* (1707) and in English as *Astronomical Lectures Read in the Public Schools at Cambridge* in 1715. They include a collection of astronomical tables taken from Flamsteed (corrected), Halley, Cassini, and Streete.

[30]Letter from Flamsteed to Sharp dated 1 July 1707; edition by Forbes, Murdin, and Willmoth (2002), p. 428.

Fig. 3.1 Whiston's
calculation of the solar
eclipse of 24 November 29
A.D. (Whiston, *Six
Dissertations*, p. 136)
(Courtesy John Hay Library,
Brown University Library)

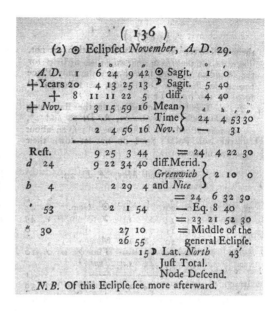

above the time of mean conjunction. For the sun he obtained a longitude of Sagittarius 1°0′and for the moon Sagittarius 5°40′, making a difference in position between the moon and sun at the moment of mean conjunction 4°40′. Finally, Whiston calculates the time it takes for the moon to travel this distance of 4°40′ using the table of the moon's mean motion on page 28 and adds to it a correction for the equation of time taken from pages 4 to 5, obtaining, with rounding along the way, a correction to the time of mean conjunction of −8h40′ (which he labels "Eq.") and finally obtained the time of true conjunction to be 23 November 21h52′30″ after noon, or 9h52′30″ in the morning on 24 November.

Whiston's final calculated time is very close to his earlier approximation of the ninth hour of the day, but somewhat earlier than that obtained by Halley and the other astronomers Sykes discussed. Whiston explains that this discrepancy is because he alone has made a correction for the moon's acceleration:

In Dr. *Halley*'s Calculation of the Eclipse *A. D.* 29. I find no Allowance made for the *Acceleration* of the Moons mean Motion, since the oldest Observations of Ecclipses (sic): For which, in all the foregoing Calculations of the first Century I have allowed 31′ and in those of the Second Century 29′ and these, as near as ever I could learn, in Agreement with the Doctors own Determination.[31]

He remarks concerning the secular acceleration that "This I have all along esteemed one of Dr. *Halley*'s greatest Discoveries in *Astronomy*" and quotes Halley's announcement of his discovery in the *Philosophical Transactions* for 1695 and Newton's remarks from the second edition of the *Principia*. Whiston's reference to the necessity of making a correction to take into account the secular acceleration when calculating eclipses in antiquity was the first published attempt to make use of Halley's discovery. Halley himself, as Whiston remarked, did not make allowance for the secular acceleration in calculating ancient eclipses, and the secular acceleration does not feature in Halley's lunar tables (long completed but still unpublished at this time). Furthermore, Whiston was the first author to quantify the effect of the acceleration:

Now my Allowance for this Acceleration of 1′ in three Periods of Eclipses or in 54 Years, must still cause about half an Hours Difference between me and others, who follow the modern Tables, in the ancient Eclipses of the First Century; and so proportionably in all other ancient Eclipses whatsoever. Which I always place so much *earlier* than the best of our present Tables determine: And, as I thought, upon Dr. *Halley*'s own Authority also.

---

[31] Whiston, *Six Dissertations*, p. 157.

I say *earlier*, because if the Moon's mean Motion be *accelerated* since the ancient Eclipses, the Month is now *shorter* than it was in old Times; tho' in a prodigiously small Degree; and so the same Number of Months reckoned backwards from our New and Full Moons including both our shorter and those longer Months, must reach *higher* from our Times, and imply the coming of Eclipses both of the Sun and Moon, to have been *earlier* than our Modern Astronomical Tables determine. This Allowance I have made in the forgoing State of the Eclipse *A.D.* 29. as from Dr. *Halley*: And shall make it in other old Eclipses, till I am satisfied that it is a Mistake. Which if it be, it behoves Dr. *Halley* to clear this Matter to the publick. It depending upon his own Observations and Assertions, and on them alone.[32]

Whiston continued by claiming that this correction is confirmed by Thucydides's account of a solar eclipse at Athens that took place during the first year of the Peloponnesian War, which without the correction for the Moon's acceleration would take place during the hours of night.

Whiston provided no explanation for how he had determined the magnitude of the time correction to account for the acceleration of the Moon to be $1'$ every 54 years and that the appropriate allowances to be made for the acceleration are $31'$ for dates during the first century B.C. and $29'$ for dates during the second century B.C., other than first to claim that the $31'$ and $29'$ corrections are "as near as ever I could learn in Agreement to the Doctors (i.e., Halley's) own Determination", and secondly that these allowances depend upon Halley's "Observations and Assertions, and on them alone". There are several issues here that require further comment.

First, Whiston made an elementary error in his treatment of the acceleration. Whiston claimed that the "allowance for this Acceleration" is $1'$ in 54 years. Assuming that the acceleration is constant, then the correction to the moon's position should increase with the square of the time, and hence the correction for the time of the eclipse should also increase quadratically, but Whiston took the allowance to increase linearly: the allowance is simply given as the number of years before the epoch of the calculations multiplied by $1'$ divided by 54 years. Whiston did not specify the epoch, but taking it for convenience to be the year of publication, 1734, the allowance in the middle of the first century A.D. is equal to $1684 \times 1/54 \approx 31'$ and in the second century A.D. $1584 \times 1/54 \approx 29'$. In a separate dissertation on various astronomical and chronological propositions included in the *Six Dissertations* Whiston applied the same formula to obtain corrections for his calculations of various ancient eclipses (see Table 3.1). For example, Whiston discussed Gaubil's dating of a reference to an eclipse in China to 2155 B.C. taking the correction to the time on account of the acceleration to be 1h11', again assuming a linear increase in the correction. Whiston calculated that this eclipse finished more than an hour before sunrise in China and redated it to 22 October 2137 B.C. As I will discuss in Chap. 7, Tobias Mayer also questioned Gaubil's dating of this eclipse.

Whiston's claim that Halley is the authority for the estimate of the size of the correction also deserves comment. Halley himself never published this estimate and so we must ask how Whiston obtained it. In his *Memoirs* Whiston portrays Halley as his friend and supporter, who was willing to second Whiston's nomination to the Royal Society.[33] Their relationship may not have been as close as Whiston presents it, however. Although Halley supported Whiston in his petition for the creation of the longitude prize, they differed in their attitudes to religion,[34] and at least at times Halley appears to have been suspicious of Whiston. A letter from the publisher and astronomer John Senex to Halley sent on 16 May 1715 illustrates Halley's attitude:

I am very much concerned that I was not at Home time enough to see you at Childs Coffeehouse, as you ordered I should the Evening before you were out of Town: what ever your orders were if you [illegible word] to communicate them by Letter, or otherwise, I will punctualy obey them. I understand from Dr Jones that you were suspicious I should [illegible word crossed out] do something for Dr Whiston…[35]

---

[32] Whiston, *Six Dissertations*, pp. 158–159.

[33] Whiston, *Memoirs*, p. 293.

[34] See Whiston's comments on a conversation with Halley in his *Memoirs*, p. 243.

[35] MS RGO 2/16 unnumbered folio.

Table 3.1 Corrections to the times of eclipses applied by Whiston to take into account the secular acceleration of the moon

Eclipse	Time correction (′)	Page in *Six Dissertations*
21 October 2137 B.C.[a]	−1,11	200
24 June 791 B.C.[a]	−47	224
8 November 771 B.C.[a]	−47	225
4 May 770 B.C.	−47	226
13 March 711 B.C.	−46	249
12 February 635 B.C.	−44	242
28 April 612 B.C.	−44	250
28 May 585 B.C.[a]	−43	229
3 May 584 B.C.[a]	−43	230
25 June 540 B.C.	−43	251
18 April 480 B.C.[a]	−40	166
10 June 437 B.C.	−40	243
20 September 331 B.C.	−39	252
7 September 253 B.C.	−37	253
21 June 168 B.C.	−35	255
26 October 63 B.C.	−33	256
30 January 30 B.C.	−33	257
30 May 29 A.D.	−31	135
24 November 29 A.D.	−31	136
20 May 30 A.D.	−31	136
13 November 30 A.D.	−31	137
9 May 31 A.D.	−31	138
2 November 31 A.D.	−31	138
28 April 32 A.D.	−31	139
22 October 32 A.D.	−31	140
22 September 32 A.D.	−31	140
12 July 120 A.D.	−29	146
6 January 121 A.D.	−29	147
2 July 121 A.D.	−29	147
26 December 121 A.D.	−29	148
21 June 122 A.D.	−29	148
7 June 476 A.D.	−25	259
15 March 1736 A.D.	0	260
15 September 1736 A.D.	0	260

[a]The printed text mistakenly gives an A.D. date

Despite Whiston's uncompromising directness in his criticism of Halley for not applying a correction to make allowance for the acceleration of the moon when calculating ancient eclipses, he is unlikely to have attempted to misrepresent the much more powerful Halley. Indeed, Halley owned a copy of Whiston's *Six Dissertations*,[36] and we can imagine that he would certainly have objected if he felt that Whiston was falsely attributing to Halley views that he did not hold. It is therefore extremely unlikely that Whiston simply made up the claim that his estimate of the moon's acceleration was in agreement with that of Halley. Two other possibilities remain. First, Halley may have fully worked out an estimate of the time correction necessary because of the moon's acceleration as 1′ in 54 years and (wrongly) assumed, like Whiston, that the correction that must be applied increases linearly with

---

[36]A copy of Whiston's *Six Dissertations* is listed in the sale catalogue of Halley's books; see Feisenberger (1975).

time. Alternatively, Whiston and Halley may have both reached similar estimates for the necessary correction in the first and second centuries A.D. of 31′ and 29′ respectively, based upon analysis of the eclipse records in Ptolemy, but only Whiston then drew the wrong conclusion that this corresponded to 1′ per 54 years. I believe that this second interpretation is the more likely. Whiston, having obtained his values of the correction for the first and second centuries and discovering that Halley had found similar values, probably then made the incorrect deduction of the magnitude and effect of the acceleration.

Finally, there is the type of correction Whiston applied. As we will see, other authors always worked with a secular equation—a correction to the longitude of the moon. Whiston's correction, however, is a correction in time, not longitude, and is applied in the calculation of the moment of syzygy. A correction of this form would be very useful for the calculation of the time of an eclipse, which is what Whiston was interested in, but of no use in calculating the longitude of the moon at an arbitrary moment. It is doubtful that Halley would have considered the secular acceleration of the moon in terms of a correction to the time of syzygy, especially considering that he used the normal longitude secular equation within his planetary tables. Again, this points to the correction to the time of syzygy of 1′ per 54 years described by Whiston as being Whiston's own work.

Whiston invoked the secular acceleration of the moon not primarily to produce better calculations of the circumstances of ancient eclipses but rather to discredit the calculations of other astronomers and support his argument that the eclipse reported by Phlegon was a supernatural event that caused the darkening of the sky at the crucifixion, not a natural eclipse. His argument, in essence, was that once he had corrected the time of the eclipse of A.D. 29 to take into account the secular acceleration, the calculated time did not agree with that reported by Phlegon, and so this eclipse could not have been the one Phlegon described. Whiston's hurry to respond to Sykes and his grasping at the secular acceleration as the means to defeat Sykes's arguments may explain why he made his mistakes in his explanation of the cumulative effect of the acceleration. Reading Whiston's discussion one has the feeling that it was written quickly in the heat of the argument, without careful thought or checking, something that Whiston was criticised for elsewhere.[37] Whiston never returned to the issue of the secular acceleration in any of his later works.

Sykes does not appear to have responded to Whiston's arguments. Whiston had not been the only person to reply to the publication of Sykes *Defence*. Almost immediately an anonymous letter to Sykes sent from Oxford was published entitled *Phlegon's Testimony Shewn to Relate to the Darkness Which happened at our Saviour's Passion*, followed by a treatise by John Chapman entitled *Phlegon Examined Critically and Impartially In Answer to the late Dissertation and Defence of Dr. Sykes*. Neither discussed the astronomical data relating to the eclipse (nor, in truth, added anything significant to the debate surrounding the eclipse). Sykes's final publication on the subject, *A Second Defence*, responded only to these two authors and did not mention Whiston.[38] It is unlikely that Sykes would have found Whiston's arguments against the identification of the eclipse with that of 25 November 29 A.D. convincing, however. Sykes had already dismissed as unimportant the difference

---

[37] The Scriblerians in their ridiculing of *Whiston* quote a contemporary of Whiston at Cambridge: "I remember Whiston well enough at Cambridge; we esteem'd him a Man of *pretty good* Learning, especially in *Mathematics*: and he held a *Correspondence* with many of our *B—ps*, who, I wonder, did not *prefer* him; the *Disappointment* of which I take to be the Cause of these his *Proceedings*: for he was always esteem'd very *ambitious*, and *self-opinionated*, of a *hot* Head and *warm* Imagination". (Scriblerius, *Whistoneutes*, p. 3). See also Buchwald and Feingold (forthcoming), Chap. 8. They quote a statement from Whiston's enemy Anthony Collins that Whison's temper "disposes him to receive any sudden thoughts, any thing that strikes his imagination, when favourable to his preconceiv'd scheme of things ... his imagination is so strong and lively on these occasions that he sometimes even supposes facts, and builds upon these facts".

[38] Sykes's *Second Defence* prompted a reply from John Chapman, *Phlegon Re-examined*, which again does not discuss the astronomical data.

in the time of the eclipse reported by Phlegon and Whiston's earlier calculation and would almost certainly have done the same with the later calculation, which did not produce a much larger discrepancy, and Whiston had no other new arguments.

Forgotten History

Whiston's discussion of the secular acceleration and his estimate for the magnitude of the resulting time correction necessary to calcule ancient eclipses—the first published estimate of a correction for the acceleration—passed almost completely unnoticed by his contemporaries and by later astronomers. Neither Dunthorne nor Mayer appear to have known of Whiston's work, and Lalande clearly states that after Halley's discovery of the acceleration, it remained for Dunthorne and Mayer to investigate it fully.[39] Most later authors clearly base their discussion on Lalande and knew nothing of Whiston's work: Roger Long in Book 3 of his *Astronomy* published in 1764 mentions only Halley and Dunthorne (and Struyck's rejection of Halley's discovery),[40] Jean-Baptiste Delambre in his *Histoire de l'Astronomie au Dix Huitème Siècle* (1827) repeats Lalande's account almost verbatim,[41] and Robert Grant in his *History of Physical Astronomy* (1852) gives what becomes the standard account of the history of the discovery of the secular acceleration: "(Halley) first alluded to this phenomena in 1963, but no attempt was made to confirm his suspicion until the year 1749, when Dunthorne communicated a memoir to the Royal Society, in which he discussed all the observations calculated to throw light on the subject".[42] To my knowledge, the only reference to Whiston's study of the secular acceleration appears in a paper published in the *Philosophical Transactions* for 1753 by George Costard concerning the date of the eclipse of Thales. Describing his calculation of the path of the eclipse, Costard wrote:

> I shall only add, that, if any allowance is to be made for the moon's acceleration, or any other cause, the track here given, as you know, will be a little different. As I cannot make several ancient eclipses that I have tried, succeed to my mind, without some such supposition, I have done the same with regard to this. What the quantity to be allowed is, I leave to you and others to determine: At present I make it about 45'; at Mr. Whiston's rate of 1' in 54 years, or thereabouts.[43]

Costard published several papers and books on the history of astronomy. Like Whiston, he was interested in the use of ancient eclipses for the study of classical and biblical chronology, and was clearly familiar with Whiston's *Six Dissertations*. Nevertheless, it is puzzling that Costard did not refer to Dunthorne's 1749 estimate of the magnitude of the secular acceleration. Perhaps because Whiston's discussion of the secular acceleration appeared in the context of studying ancient eclipses for the purpose of chronology, whereas Dunthorne's paper is concerned purely with lunar theory, Whiston's result appealed more to (or at least was read by) a scholar interested in chronological issues.

Nevertheless, it is strange that Whiston's discussion of the secular acceleration passed completely out of history. Even Nicolaas Struyck who as I will discuss in the next chapter argued in 1740 against the existence of the secular acceleration did not mention Whiston's discussion of the topic despite referring to other parts of Whiston's *Six Dissertations* in which this discussion was published. The reasons

---

[39] Lalande, "Mémoire sur les Équations Séculaires", pp. 426–472.

[40] Long, *Astronomy*, pp. 433–434.

[41] Delambre (1827), pp. 597–598.

[42] Grant (1852), p. 60.

[43] Costard, "Concerning the Year of the Eclipse foretold by Thales", pp. 24–25.

why Whiston's contribution was ignored and forgotten can only be speculated. It may be that other astronomers saw the error in Whiston's interpretation of the effect of the acceleration over time and so disregarded his discussion of the topic. I suspect a more likely reason is simply that Whiston's discussion was presented within a polemical debate that did not interest a large audience, and so was scarcely noticed and quickly forgotten. In the early 1740s Thomas Birch wrote in his collection of historical and biographical notes an account of the debate based in large part upon information he had been given by Sykes which he begins "The question whether <u>Phlegon</u> spoke of the Darkness at the Time of the Passion of our lord was canvassed ere about 8 years ago in several Dissertations proteon".[44] Birch's account presents the episode as a short but intense debate lasting only a couple of years. It is quite likely, therefore, that Dunthorne and other astronomers of the 1740s and 1750s were completely unaware of Whiston's discussion of the secular acceleration. Whiston himself never discussed the subject again. One wonders if he himself recognized the mistake in his analysis of the effect of the acceleration over time and so did not wish to revisit the issue where he would have been forced to correct his error. More likely, however, is that the secular acceleration had served its purpose in providing ammunition in his debate with Sykes, after which it was no longer relevant in his future work.

---

[44]BL MS Add 4224, f. 220. The account spans ff. 218–222 and includes a fragment of Sykes' original draft for the end of his *Second Defence* concerning the letter of the year of the eclipse that Sykes decided to omit from the published version.

Chapter 4
The Gradual Acceptance of the Existence of the Secular Acceleration During the 1740s

> *For it seems confirmed by undeniable proofs, that the moon absolves her period in a shorted compass of time than formerly.*
>
> —George Smith, *A Dissertation on the General Properties of Eclipses* (London 1748), p. 15.

The 1740s saw a major shift in the way that solar, lunar, and planetary theories were formulated. Continental mathematicians applied the calculus as interpreted by Leibniz in an attempt to solve the three-body problem of the gravitation forces in the Earth–moon–sun system to compute the perturbations in the lunar orbit, and eventually extended this procedure to the planets. Celestial mechanics shifted from being an astronomical problem to a mathematical one, and the locus of its study moved from England to continental Europe and from astronomers to mathematicians. Consideration of the mechanics of the orbits of the bodies in the solar system led mathematicians such as Leonard Euler back to the question of whether the heavenly bodies were retarded in their motion by the æther. The secular accelerations of the sun, moon and planets therefore took on a new significance and by the end of the 1740s the existence of these accelerations was generally accepted, although the size of the accelerations remained unsolved. The decade began, however, with a challenge to Halley's discovery of the lunar secular acceleration.

An Argument Against Halley's Discovery: Nicolaas Struyck

Nicolaas Struyck (1687–1769), a Dutch astronomer, mathematician, teacher, and author of important works on geography and the statistics of populations,[1] rejected Halley's discovery of the secular acceleration of the moon in his *Inleiding tot de Algemeene Geographie* of 1740:

> Mr. Halley believed that one can conclude, from the Babylonian lunar eclipses, compared with those of Albategnius and today, that the Moon has begun to move faster than in earlier times. In order to see if that is true, I shall give a short excerpt from a research I have made concerning this. Almost all of the prominent astronomers have calculated the nineteen ancient lunar eclipses that Ptolemy has transmitted, and nevertheless one cannot bring them closely enough into agreement with the tables. That does not surprise me. Apart from the obvious contradictions that were already discovered in the texts of Ptolemy by Bullialdus and others, I have found still

---

[1] For brief biographical details of Struyck, see Hald (2003), pp. 394–396.

J.M. Steele, *Ancient Astronomical Observations and the Study of the Moon's Motion (1691–1757)*, Sources and Studies in the History of Mathematics and Physical Sciences, DOI 10.1007/978-1-4614-2149-8_4, © Springer Science+Business Media, LLC 2012

Table 4.1 Struyck's analysis of the lunar eclipses reported by Ptolemy

Date of eclipse	Difference in longitude between observation and calculation		
	Struyck	Streete	Bullialdus
19 March 721 B.C.	–	[−12′57″]	[−4′54″]
8 March 720 B.C.	+1′56″	+11′9″	+0′42″
1 September 720 B.C.	–	[−28′11″]	[−19′24″]
21 April 621 B.C.	−11′50″	+13′53″	+6′12″
16 July 523 B.C.	−5′20″	−30′6″	−26′50″
19 November 502 B.C.	−1′17″	+6′17″	−10′12″[a]
25 April 491 B.C.	+0′54″	−6′4″	+19′56″
22 December 382 B.C.	+9′54″	−12′28″	−1′26″
18 June 382 B.C.	+0′21″	−13′54″	−6′15″
12 December 382 B.C.	+0′58″	−0′23″	−6′52″
22 September 201 B.C.	−3′38″	+8′27″	−10′9″
19 March 200 B.C.	−1′34″	+11′35″	+19′7″
11 September 200 B.C.	+4′10″	−15′7″	−10′5″
30 April 174 B.C.	+5′24″	+0′41″	+18′26″
27 January 141 B.C.	+1′24″	+3′30″	+18′39″
5 April 125 A.D.	−0′25″	−10′58″	−14′15″
6 May 133 A.D.	+0′21″	+8′47″	+2′10″
20 October 134 A.D.	−2′25″	+2′11″	−2′6″
5 March 136 A.D.	+2′43″	−25′57″	+35′6″

In addition to his own calculation of the difference, Struyck gives values
from Streete's *Astronomia Carolina* and Bullialdus's *Astronomia
Philolaica*. Struyck does not give the data for the eclipses of 19 March
721 B.C. and 1 September 720 B.C.; in Table 4.1 have given Streete's
and Bullialdus's values for these eclipses from the original works
[a]Bullialdus actually gives −18′12″

other difficulties. When I made calculations, according to the tables which presented reasonably well the eclipses
both old and new, I saw that, out of three lunar eclipses occurring within a year and a half of each other, that two
would agree well with the calculations, but the third would sometimes differ by as much as a half-hour or even
an hour. This can be attributed not to the Moon's average course, but to the dissimilarities upon which these
(tables) are dependent.[2]

Struyck explained that the times of several of the eclipses reported in the *Almagest* were incorrectly
identified by Ptolemy as referring to the beginning rather than the middle of the eclipse. In addition,
Struyck believed that some of the eclipses were observed in Athens, not Babylon or Alexandria as
Ptolemy had reported. Comparing the differences between the longitude of the moon at the middle of
the eclipses derived from his re-interpretations of the accounts given by Ptolemy with calculated longitudes taken from his own work, from Streete and from Bullialdus (see Table 4.1), Struyck found that
for his own calculations, the difference in the moon's longitude given by observation and calculation
was, with two exceptions, always less than 6′, which was within the tolerance of the precision of the
ancient timings. The times of the eclipses, therefore, provide no evidence for the existence of the secular
acceleration; instead, they suggest that the moon had *not* accelerated:

Since, in 1563 years, or 19332 lunations, during which each of these eclipses occurred, no difference of any
importance can be found in the average course, I believe that one can easily conclude that the Moon moves neither
any faster nor any slower than in earlier times.[3]

---

[2] Struyck, *Inleiding tot de Algemeene Geographie*, p. 81. This and the following quotation from Struyck are translated
from the Dutch by W. S. Monroe.

[3] Struyck, *Inleiding tot de Algemeene Geographie*, p. 85.

Why did Struyck's calculations fit the circumstances of the eclipses better than those of Streete or Bullialdus? Partly it was because Struyck used different values for the longitude of Babylon and Alexandria, values that were not necessarily more accurate but which placed the cities slightly further west than previous estimates. The major factor, however, was Struyck's assumption that the seven eclipses dating between 383 and 174 B.C. were observed in Athens.[4] Ptolemy expressly says, however, that the eclipses of 23 December 383 B.C., 18 June 382 B.C., and 12 December 382 B.C. were brought over from Babylon by Hipparchus,[5] and that the eclipses of 22 September 201 B.C., 20 March 200 B.C., 12 September 200 B.C., and 1 May 174 B.C. were observed in Alexandria.[6] But by moving the eclipses to Athens, Struyck significantly reduced the discrepancy between the accounts of the observations and calculation, removing the need for the secular acceleration.

Struyck's rejection of the existence of the moon's secular acceleration made very little impact on contemporary astronomers. Only Roger Long, in his second part of his *Astronomy* published in 1764, commented upon it, and then in unfavourable terms:

> *Nicolaus Struyck* author of a treatise of geography and astronomy in low Dutch opposes *Halley*'s opinion of the acceleration of the moon's motion; but the shifts he makes use of to evade the force of the proofs arising from the difference between the observed and the computed places in the ancient eclipses are very extraordinary.[7]

Long explains that Hipparchus's dating of the eclipses using the name of the Athenian archon and Athenian month names is perfectly understandable for a Greek author, "as it would be for a Roman historian writing in Latin to distinguish the time of any event which fell out in a distant country by saying who were consuls at Rome that year".[8] It cannot be taken to indicate that the eclipses were observed in Athens. Similarly, Long says that the use of the Callipic calendar in the Alexandrian observations does not imply that they were actually observed in Athens.

Leonard Euler

Despite Struyck's attempt at a rejection of Halley's discovery of the moon's secular acceleration, throughout 1740s the idea that the motions of the heavenly bodies were subject to long-term changes was gaining wider acceptance. In 1746 Leonard Euler gave a detailed mathematical argument for the gradual increase in the mean velocity of the sun (or rather the Earth) and the planets because of the resistance of the æther to their motion.[9] Euler was by this time one of the leading mathematicians in Europe. Born in Basel in 1707, Euler studied under Johann Bernoulli, who feted him as a man with exceptional mathematical talents, and had held positions at the Academy of St Petersburg and the Berlin Academy before reaching his mid-thirties.[10] Euler had been one of the first to apply trigonometric functions to the solution of differential equations and during the mid-1740s was applying this to the problem of solar and lunar motion.[11]

---

[4] The idea that some or all of the Babylonian eclipses of 383–382 B.C. were observed in Athens was revived by Oppolzer (1881) and Nevill (1906); see van der Waerden (1958) for a comprehensive criticism of the claim.

[5] *Almagest* IV.11; Toomer (1984), p. 211.

[6] *Almagest* IV.11 and VI 5; Toomer (1984), pp. 214 and 283.

[7] Long, *Astronomy*, p. 433.

[8] Long, *Astronomy*, p. 433.

[9] For a detailed study of this topic, see Wilson (1980), pp. 93–104.

[10] For a recent biography of Euler, see Fellmann (2007).

[11] Wilson (2010), pp. 10–13.

On 15 February 1746 Euler wrote to Delisle announcing a forthcoming volume of dissertations. These would include his new theory of light and colour as well as his discovery that the length of the year was not constant. Euler's discussion duly appeared with the title "De relaxtione motus planetarum", published in the first volume of his *Opuscula varii argumenti* (1746). In this work Euler attempted to calculate the effect on the motion of bodies in the solar system through the æther, whose density he determined from the velocity of light.[12] He concluded that the period of the Earth's orbit diminishes over time. In his letter to Delisle, Euler remarked that this is consistent with observation which had shown that the current year length was shorter than that found in Cassini's tables and from observations from the previous three centuries. The ancient observations reported by Ptolemy appeared to show the opposite, but Euler suggested that this was because of errors of a day or so made in the conversions of Ptolemy's data to the Julian calendar.[13]

In the *Opuscula varii argumenti* Euler also published a new set of solar and lunar tables. He included a table for the correction to the sun's longitude to take into account a secular acceleration. Euler took the coefficient of the secular equation of the sun to be $1°$ in 1900 years. In a letter to Caspar Wetstein, chaplain to the Prince of Wales, dated 28 June 1749 and later published in the *Philosophical Transactions*, Euler explained that he used observations by Bernard Walther to examine the acceleration of the sun, and that he hoped that earlier observations will come to light that would confirm his findings.[14] In particular, Euler expressed hope that Arabic observations from the work of Ibn Yūnus will soon be available to him. He had heard from Lemmonnier of the existence of a manuscript of Ibn Yūnus in Leiden (Euler initially thought that manuscript was in Oxford, but communicated a correction before the letter was published),[15] which must have been the manuscript that had belonged to Golius from which Curtius had obtained translations of three eclipse records (see Chap. 5). On 20 December Euler wrote again to Wetstein explaining in more detail the problems with using Ptolemy's observations because of possible errors in the day count (an argument which, as we will see, Mayer later convinced Euler was incorrect). The observations of Ibn Yūnus, however, would not suffer the same problem "because the Julian Kalendar has not been interrupted for these last past twelve hundred Years",[16] a not-very-convincing argument. Euler continues by referring to Halley's discovery of the secular acceleration of the moon: "The late Dr. *Halley* had also remark'd that the Revolutions of the Moon are quicker at present than they were in the Time of the ancient *Chaldeans* who have left us some Observations of Eclipses".[17] Rather than taking this as further evidence for the resistance of motion of heavenly bodies through the æther, as might be expected, Euler instead raises a new question: is the change in orbital period of the Moon real or could it be an apparent effect caused by a change in the rate of rotation of the Earth:

> But as we measure the Lengths of Years by the Number of Days, and Parts of a Day, which are contained in each of them; it is a new Question. Whether the Days, or the Revolutions of the Earth round its Axis, have always been of the same Length. This is unanimously supposed, without our being able to produce the least Proof of it.[18]

If the length of day was different today from what it had been in antiquity, then since any ancient observations would be timed using this different day length, this could cause the apparent speed of the Moon to differ between antiquity and the present.

---

[12] For a detailed presentation of Euler's mathematical arguments, see Wilson (1980), pp. 93–104 and Aiton (1996), pp. li–lviii.

[13] Kepler had previously suggested calendrical irregularities as the reason for the errors in Ptolemy's equinox observations in book 7 of the *Epitome Astronomiæ Copernicanæ*. See Swerdlow (2010), pp. 190–197.

[14] Euler, "Concerning the Gradual Approach of the Earth to the Sun".

[15] Aiton (1996), p. 257, fn. 2.

[16] Euler, "Concerning the Contraction of the Orbits of the Planets", p. 357.

[17] Euler, "Concerning the Contraction of the Orbits of the Planets", pp. 357–258.

[18] Euler, "Concerning the Contraction of the Orbits of the Planets", p. 358.

The General Acceptance of the Existence of the Secular Accelerations

George Smith, an English schoolmaster, amateur astronomer, and frequent contributor to the *Gentleman's Magazine* on topics ranging from geology, medicine, and botany to the prehistoric and Roman antiquities and topography of Cumbria,[19] presented the moon's secular acceleration as an established fact in his *Dissertation on the General Properties of Eclipses* (1748). In this short pamphlet, Smith discussed the application of eclipse periods to the prediction of eclipses, in particular the 18-year Saros cycle which he calls the "Chaldean period", the link between biblical prophecy and eclipses, and presented a detailed map of the path of the total solar eclipse of 24 July 1748 with illustrations of the extent of the eclipse as it would be seen in various important cities. Smith ended his pamphlet by mentioning the effect of the moon's secular acceleration on the calculation of eclipses:

> *P.S.* It is particularly to be noted, that eclipses, which have happened many centuries ago, will not be found by our present tables to quadrate exactly with ancient observations, by reason of the great anomalies in the lunar motions; which appears an incontestable demonstration of the non-eternity of the universe. For it seems confirmed by undeniable proofs, that the moon absolves her period in a shorted compass of time than formerly, and will continue by the centripetal law to accelerate in velocity as she approaches the earth; nor will the centrifugal power be found sufficient to compensate the different gravitations of such an assemblage of bodies as constitute this mighty fabric, which would rush to ruin itself, without proper assistance to regulate and adjust to its original motions.[20]

Smith's explanation that the moon's motion is accelerated because it is gradually falling towards the sun is similar to the arguments made by Halley in his unpublished paper read to the Royal Society on 18 October 1683 and to the comments of David Gregory in his *Elements of Astronomy*. For Smith, however, the cause of the moon's acceleration is less important than the fact of its existence, and he moves swiftly to a discussion of the eclipse of Thales and an eclipse reported by Thucydides as examples showing that the secular acceleration of the moon is necessary in order for calculation to agree with the reported observation. He continues:

> I have only obviated these things by way of caution to the present astronomers, in re-computing antient (sic) eclipses, and refer them to examine the eclipse of *Nicias* so fatal to the Athenian fleet, that which overthrew the *Macedonian* army, &c. from the annals of history, because the tedious process of calculation is foreign to my design. Mean time I shall note down 10 or 12 remarkable eclipses, that will be visible at *London* in the compass of 800 years, as they may serve for a future standard to succeeding astronomers, in ascertaining the quantity of this increased velocity of the lunar motions, for which a proper table may be composed, to regulate, with more certainty, eclipses of great antiquity; as they will have larger opportunities to gratify their knowledge, and confirm this conjecture, from the numerous registers of those phenomena daily published.[21]

For Smith, the importance of the secular acceleration was in the need to take it into account when calculating the circumstances of past and future eclipses. His hope is that future astronomers will succeed in quantifying the acceleration and be able to formulate a table to supplement existing lunar tables by providing the necessary correction to take into account the moon's increased velocity over time. Within a year Smith's desired table would be provided by Richard Dunthorne.

---

[19] de Montluzin (2001) and (2004).

[20] Smith, *A Dissertation on the General Properties of Eclipses*, pp. 15–16.

[21] Smith, *A Dissertation on the General Properties of Eclipses*, pp. 17–18.

Chapter 5
Eighteenth-Century Views of Ancient Astronomy

> It is a judicious Observation of a very eminent Historian "that it is highly difficult to
> arrive at the Truth of past Transactions; as Reports are usually transmitted from
> Hand to Hand without any one's being at the Trouble to examine them".
>
> ——George Costard, *The Use of Astronomy in History and Chronology*
> (London, 1764), p. 1.

By the end of the 1740s the existence of the moon's secular acceleration was widely accepted among astronomers. The magnitude of the acceleration, however, remained unknown, excepting William Whiston's rushed and unjustified claim that it caused of correction to the time of an eclipse in the past, which increases at rate of 1 minute in 54 years. Between 1749 and 1757 three attempts were made to determine the magnitude of the moon's secular acceleration by Richard Dunthorne, Tobais Mayer, and Jérôme Lalande after which attention shifted to trying to account for the acceleration theoretically. Dunthorne, Mayer and Lalande's study of the moon's secular acceleration relied upon the interpretation and exploitation of ancient astronomical records. How to interpret ancient astronomical observations and which sources of ancient records could be relied upon had been a controversial issue since the Renaissance and would require Dunthorne, Mayer, and Lalande to make decisions that would have a direct impact upon their estimates of the size of the secular acceleration.

Sources for Studying the Secular Acceleration of the Moon

To quantify the secular acceleration of the moon, Dunthorne, Mayer, and Lalande relied extensively upon historical records of eclipses. The cumulative effect of the secular acceleration on the longitude of the moon increases with the square of time. As a result, the imprecision of eclipse reports from antiquity is offset by their greater distance back in time: the earlier a record, therefore, the greater its potential utility in the investigation of the secular acceleration. As a consequence, Dunthorne, Mayer, and Lalande sought out ancient and medieval records of lunar and solar eclipses.

Several compilations of early records were available by the mid-eighteenth century, most prominently those by Giovanni Battista Riccioli and Albertus Curtius. Riccioli's *Almagestum Novum*, published in Bologna in 1651, contained in volume 1, pp. 361–368 a catalogue of all eclipse observations known to Riccioli from antiquity down to his own time, including the eclipses recorded in Ptolemy's

J.M. Steele, *Ancient Astronomical Observations and the Study of the Moon's Motion (1691–1757)*,
Sources and Studies in the History of Mathematics and Physical Sciences,
DOI 10.1007/978-1-4614-2149-8_5, © Springer Science+Business Media, LLC 2012

Almagest, Theon of Alexandria observation of a solar eclipse in A.D. 364 (incorrectly dated to A.D. 365 by Riccioli, followed by Curtius), eclipses seen by al-Battānī as known from the Latin translation of his work *De Scientia Stellarum*, more recent eclipse observations by Regiomontanus, Bernard Walther, and Tycho Brahe, as well as many accounts of lunar and solar eclipses in classical histories and medieval chronicles.[1] Albertus Curtius, under the pseudonym Lucius Barrettus (an anagram of his name), edited some of Tycho Brahe's astronomical observations and tables for publication in 1666 under the title *Historia Coelestis* and included a lengthy Prologomena cataloguing astronomical observations known from antiquity down to Tycho's time. This work covered mostly the same material as Riccioli's catalogue, with the important addition of three eclipse observations taken from Ibn Yūnus. The reports of these three eclipses were known from a unique Arabic manuscript of the *zīj* of Ibn Yūnus in the possession of Jacobus Golius, Professor of Arabic and of Mathematics at Leiden. After several failed requests from English astronomers and Arabists asking for details of any astronomical observations contained in this manuscript, Golius eventually sent information on these three observations to Wilhelm Schickard, a professor in Tübingen, from whose manuscript notes Curtius obtained a Latin translation of the reports of the three eclipses.[2] Either of these compilations could have provided a convenient starting point for astronomers seeking historical eclipse observations.

The method of determining the size of the secular acceleration of the moon employed by Dunthorne and his successors relied upon comparing the longitude of the moon deduced from an eclipse observation with the moon's longitude calculated for that moment using current lunar theory based upon an unchanging mean velocity. The secular acceleration could then easily be calculated from the difference in the longitudes and the number of years between the observation and the epoch of the lunar theory. Only certain historical reports of eclipses were suitable for this analysis. First, their date and place of observation must be confidently known. This immediately ruled out most of the eclipses reported in the works of classical historians and from medieval chronicles, which were usually ambiguous in one if not both of these details. Secondly, the time of day or night at which the eclipse was seen was required. Again, few eclipses reported by historians or chroniclers included this information, which was necessary when calculating the longitude of the moon using tables. Only eclipses observed and recorded by astronomers fulfilled these requirements, which effectively reduced the sources available for investigating the secular acceleration to the 19 well-known eclipses in Ptolemy's *Almagest*, the solar eclipse observed by Theon of Alexandria, the 3 eclipses reported by Ibn Yūnus given in Curtius' *Historia Coelestis*, and the eclipses in al-Battānī's *De Scientia Stellarum*, plus several eclipses dating from the fifteenth and sixteenth century observed by Regiomontanus, Walther, Copernicus, and Tycho.

One other source of historical observations became available to astronomers in the late seventeenth and early eighteenth centuries: China. In the late sixteenth century European Jesuits began a mission to China that would result in their acting as a conduit for European astronomy to become established in China, and for knowledge of Chinese history, literature and astronomy to enter Europe. Science, and especially astronomy, was used as a gateway into the court of the Chinese Emperor by the Jesuits; as a result, several of the Jesuits who travelled to China were accomplished mathematicians and astronomers with links into the community of scholars in Europe. Their work culminated with the adoption by the Qing dynasty in 1644 of an official calendar (*li* 曆), the *Shixianli* 時憲曆 "Temporal Pattern System", compiled by the Jesuit Adam Schall von Bell.[3] In the second half of the seventeenth century, some returning Jesuits published accounts of China and its history, at least two of which,

---

[1] Riccioli published a similar catalogue in his *Astronomia reformata* of 1665, vol. 2, pp. 95–104 and 143–147, but mistakenly changed the dates of the eclipses of al-Battānī.

[2] Toomer (1996), p. 48.

[3] See Sivin (2011) for a discussion of the imperial significance of the calendar in China and details of the 1644 reform.

Martino Martini's *Sinicae Historiae Decas Prima* (1659) and Phillipe Couplet's *Tabula Chronologica Monarchiae Sinicae* (1686), contained reports of very ancient Chinese astronomical observations. These works attracted the attention of the French astronomer Joseph-Nicolas Delisle among others.[4] During the first half of the eighteenth century the Jesuit Antoine Gaubil communicated from Beijing extensive reports on Chinese astronomy to the Royal Society and other scientific bodies, wrote several treatises on Chinese astronomy and chronology that far surpassed any that had come before, and corresponded with many astronomers in Europe. By the time Dunthorne published his paper on the secular acceleration, the 36 solar eclipses recorded in the *Chunqiu* 春秋 dating from the eighth to the fifth century B.C. were well known along with several extremely ancient eclipses believed to date to the third millennium B.C. However, none of these records from China were used by Dunthorne, Mayer, or Lalande in their attempts to determine the secular acceleration.

Writing the History of Astronomy in the Eighteenth Century

In order to understand Dunthorne, Mayer, and Lalande's use of historical sources, and in particular their choices of which sources to give the most weight to, it will be helpful to investigate the descriptions of astronomy in ancient cultures presented in contemporary histories of astronomy. Eighteenth-century histories of astronomy were written in two genres: historical overviews included in books of astronomy (frequently as extensive Prologomena, though sometimes as the concluding chapters of the work), and individual books or papers devoted solely to the history of astronomy. Both genres were well established already in the sixteenth and seventeenth century.[5] For example, Bullialdus's *Astronomia Philolaica* (1645) and Riccioli's *Almagestum Novum* (1651), both works ostensibly concerning contemporary astronomy, contain extensive discussions of the history of astronomy. Works written purely as histories of science include Bernardino Baldi's *Vite dei Mathematici* (1587–1595) and Cassini's *De l'Origine et du Progre's de l'Astronomie* (1693). Eighteenth-century histories of astronomy can be seen as continuing these two traditions.

Historical writing about astronomy was fairly widespread during the eighteenth century. Many of the leading astronomers of the period, including John Flamsteed, Nicholas Louis De Lacaille, Pierre-Simon Laplace and, most notably, Jean Baptiste Joseph Delambre wrote detailed treatments of the history of astronomy from its origins down to their own times, whilst lesser known figures such as Pierre Estève and George Costard published books and papers that were no less influential at the time. Dunthorne, Mayer, and Lalande themselves each wrote about the history of astronomy. Dunthorne and Lalande made their contributions to the history of astronomy later in life, several years after their work on the secular acceleration (in Dunthorne's case this work was incorporated into the final book of the late Roger Long's *Astronomy*, which Dunthorne and subsequently William Wales saw to completion). Mayer's historical writing was confined to his unpublished lectures and his correspondence with Euler and other scientists. I will treat the historical writings of these men in the following chapters.

The first major history of the astronomy published in the eighteenth century appeared as an extensive preface to the third volume of Flamsteed's *Historia Coelestis Britannica* of 1725.[6] Flamsteed, the

---

[4] Hsia (2008).

[5] A general overview of sixteenth- to eighteenth-century works on the history of science is given in Zhmud (2006), pp. 1–10. See also Goulding (2006), Popper (2006), and Swerdlow (1993).

[6] An English translation of Flamsteed's historical preface is given in Chapman and Johnson (1982), which also includes a discussion of the writing of the preface and its purpose.

first Astronomer Royal, began working on the publication of the *Historia Coelestis Britannica* in the first few years of the eighteenth century and by 1704 had substantial parts of it ready for the press. The work was to contain Flamsteed's observations made at Derby and Greenwich and a newly created "British Star Catalogue". The publication of the *Historia Coelestis Britannica* was beset by problems from the beginning, however.[7] Printing problems, lack of money to pay Flamsteed's assistants, and a growing acrimony between Newton and Flamsteed resulted in the eventual collapse of the project in 1712 when Newton and Halley published an unauthorized version of the work as *Historial Coelestis Libri Duo*. Flamsteed was outraged, not only that the work had been published without his consent and contained open attacks on Flamsteed in its preface[8] but also because of the large number of errors the work contained. In 1714 Flamsteed succeeded in buying up around 300 unsold copies of the bootleg edition, stripped them of the few pages he felt were useful (to be inserted in his own edition when it could finally be printed) and burnt what was left. He returned to preparing his own authoritative edition, completing it in 1719, the year of his death. The *Historia Coelestis Britannica* was eventually printed in three volumes in 1725.

Flamsteed wrote his historical preface, which appeared in the third volume, both to set his work in historical context and to vindicate Flamsteed himself from the criticisms that had been levelled at him by his enemies in the Royal Society.[9] He completed the preface in the month before he died, writing it in English and apparently intended for it to be published in the vernacular rather than in Latin. However, following Flamsteed's death it was decided by his wife and his assistants, who took charge of the publication, to have the preface translated into Latin.

Flamsteed's history of astronomy is a history of observations and instruments. His focus was on the progress from crude, inaccurate observations made by the ancients to his own precise and accurate observations and his pioneering use of instrumental technology at Greenwich.[10] Astronomical theory was barely mentioned: Copernicus' work was dealt with in only half a paragraph, and that focused on the relative accuracy of the Prutenic tables that were constructed by Reinhold on the basis of Copernicus' theories as compared to the Alphonsine tables when tested against observation. By contrast, Bernard Walther received five paragraphs and Tycho Brahe a long chapter, which included extensive excerpts from Tycho's *Astronomiae Instauratae Mechanica* detailing Tycho's observational instruments. Flamsteed's discussion of Ptolemy was concerned purely with the observations reported in the *Almagest*, the instruments he used and their errors, and the *Almagest* star catalogue. Ptolemy's solar, lunar, and planetary theories, which formed the bedrock of all theoretical astronomy in the west up to the sixteenth century, was passed over in silence by Flamsteed.

The focus on astronomical observation and instrumentation in Flamsteed's preface reflects his own commitment to an observational astronomy driven towards increasing accuracy and precision of observation. The preface allowed Flamsteed to place his own work in historical context, justifying its importance by stressing the superiority of his own observations over what had went before. In doing so, he was able to challenge those critics who complained that despite all the money invested in the

---

[7] On the publication history of the *Historia Coelestis Britannica*, see Chapman and Johnson (1982), pp. 8–14.

[8] These attacks included the quite scandalous statement that "Flamsteed had now enjoyed the title of Astronomer Royal for nearly 30 years but still nothing had yet emerged from the Observatory to justify all the equipment and expense, so that he seemed, so far, only to have worked for himself or at any rate for a few of his friends..." (translation by Chapman and Johnson (1982), p. 191). Although Flamsteed, who had had to pay for most of the Observatory's instruments out of his own (frequently in arrears) salary, had published few of his observations he had sent a substantial number to Newton.

[9] Chapman and Johnson (1982), p. 1.

[10] Chapman and Johnson (1982), p. 6.

position of Astronomer Royal and the Observatory, little had been achieved.[11] Flamsteed's history of earlier astronomers must be seen in this context.

Johann Friedrich Weidler's *Historia Astronomiae sive De Ortu et Progressu Astronomiae*, published in 1741, promised to be a more thorough treatment of the history of astronomy, dealing with observations, cosmologies, and astronomical theory as well as giving biographical accounts of astronomers from the beginning of the world down to his own time (the last entry dealing with Christian Ludovic Gersten and the Giessen observatory is dated to 1740). Weidler, a professor of mathematics at Wittenberg University, made astronomical and astro-meteorological observations (mock suns, aurora borealis, etc.) during the late 1720s and the 1730s. In 1732 he was elected to the fellowship of the Royal Society, regularly communicating reports of his observations to its members.

Weidler's *Historia Astronomiae* is divided into 16 chapters covering the following topics: (1) the fabulous origins of astronomy, (2) the astronomy of the patriarchs, (3) the Chaldeans and Phoenicians, (4) the ancient Egyptians, (5) the Greeks before the founding of the "Alexandrian School", (6) the period from the Alexandrian School to the birth of Christ, (7) the first 8 centuries A.D., (8) the Arabs, (9) the Persians and the Tartars, (10) other Orientals (Mongols, Siamese, Chinese, and Americans), (11) the Jews, (12) the middle ages (ninth to fourteenth centuries A.D.), (13) the fifteenth century, (14) the sixteenth century, (15) the seventeenth century, and (16) the eighteenth century. Weidner's arrangement of history into these periods and cultures provided the basic model for several later histories of astronomy including the historical final book of Roger Long's *Astronomy* and, to a large extent, Pierre Estève's *Histoire Generale et Particuliere de l'Astronomie*, despite Estève's criticism of Weidler.

Although extremely thorough in its coverage of astronomers, listing several hundreds of individuals many of whom are forgotten today, Weidler's book provided only a very superficial description of their astronomy. His approach was primarily bio- and bibliographical, based around reporting the titles and quoting from astronomical works with very little analysis of their content or discussion of their place in the development of astronomy. Copernicus, for example, received just three longish paragraphs, extending over two and a half pages, most of which described Copernicus' upbringing and life, the printing of *De Revolutionibus* focusing on Rheticus' role in its publication, Osiander's preface, and later appraisals of the work by Gassendi, Curtius, and Bullialdus; the substance of Copernicus' astronomy was given only a summary treatment. Tycho received a fuller account, focusing on Uraniburg, his political connections, and his observations. Weidler's accounts of ancient astronomy provide further examples of his bibliographical approach to writing history. His discussion of Ptolemy takes up about eight pages of which six list Ptolemy's books, discussing the published editions and translations, but gives only the barest of details about Ptolemy's astronomy. For example, all he had to say on Ptolemy's planetary theories was that for "the inequalities of motion, at one time by means of eccentrics, at another time by means of epicycles, (Ptolemy made) elegant refinements".[12] Weidler's section on Chaldean astronomy surveys many classical sources, often with extensive quotations, but again with little attempt at critical analysis.

Weidler's book provided the basis for a short Latin history of astronomy written by Ralph Heathcote in 1747. The *Historia Astronomiæ, sive, De Ortu & Progressu Astronomiæ* presents a short account of the history of astronomy from its beginnings until Newton accompanied by a discussion of ancient methods of philosophy and that of Newton. Heathcote was born in 1721 in Leicestershire, the son of Ralph Heathcote, a curate, and Mary, the daughter of Simon Ockley, professor of Arabic at Cambridge.[13]

---

[11] To conclude the preface, Flamsteed wrote a detailed account of his battles with Newton over the publication of his work, but this section was omitted from the published edition. The text was eventually printed in Francis Baily's *An Account of the Rev. John Flamsteed* (1835). See also Chapman and Johnson (1982), pp. 160–180.

[12] Weidler, *Historia Astronomiae*, p. 178.

[13] For biographical details, see Young (2004). Heathcote's own account of his life is published in Nichols (1812), vol. III, pp. 531–540, with further biographical details by Nichols on pp. 540–544.

He was educated at Jesus College, Cambridge, becoming AB in 1745 and obtaining the MA in 1748. Recognizing that he was unlikely to obtain a fellowship, Heathcote left Cambridge to take up a career in the church and wrote several treatises on the truth of biblical miracles and matters of the Christian faith. The *Historia Astronomiæ* was his only scientific work, and "though it cannot well be considered otherwise than as a juvenile production", Heathcote later wrote, it "was yet kindly received by the University, and laid the foundation of what little merit I have since acquired in the world of letters".[14]

Heathcote's *Historia Astronomiæ* was written the year he left Cambridge. Heathcote's history follows the familiar path from astronomy's origins in the time before the deluge down through the Chaldeans and Egyptians, the Greeks, the Romans, the Arabs, to the breakthroughs of European astronomers during the Renaissance culminating in the work on Newton. In addition to Weidler's *Historia Astronomiae*, Heathcote relied on the usual mix of Josephus and classical sources for ancient astronomy. His history is unusual, however, in emphasizing the philosophical side of astronomy, discussing topics such as Aristotelian philosophy during the Middle Ages.

The year before Ralph Heathcote published his *Historia Astronomiæ*, another Englishman, George Costard, wrote *A Letter to Martin Folkes, Esq., President of the Royal Society, Concerning the Rise and Progress of Astronomy amongst the Ancients* (1746). Over the next 2 years, Costard supplemented his history with a paper on Chinese chronology and astronomy published in the *Philosophical Transactions*, and *A Further Account of the Rise and Progress of Astronomy amongst the Ancients, in Three Letters to Martin Folkes, Esq; President of the Royal Society* (1748). Costard was educated at Wadham College, Oxford, where he learnt Hebrew and Arabic and was chosen proctor of the University in 1742.[15] He served as curate of Islip, Oxfordshire, vicar of Whitchurch, Dorset, and was presented with the vicarage of Twickenham, Middlesex, in 1764. Costard was offered the wardenship of Wadham College in 1777, but declined and continued to live in Twickenham until his death in 1782, apparently dying penniless, save for a fine library of books, manuscripts, and scientific instruments. In addition to his works on the history of astronomy, Costard published studies of the dates of eclipses recorded in classical history, the Bible, oriental philology, and was an occasional contributor to the *Gentleman's Magazine*. Costard's *The History of Astronomy, With Its Application to Geography, History and Chronology, Occasionally Exemplified by Globes* (1767), despite its title, is primarily aimed at teaching students spherical astronomy and the use of a globe; the history of astronomy is used to illustrate and explain astronomical problems.

Costard's histories of astronomy are restricted to antiquity and have the explicit aim of restoring the reputation of the Greeks as the first people to make astronomy a science. He began *A Letter Concerning the Rise and Progress of Astronomy* by stressing that although the Egyptians and Babylonians are commonly said to have provided the foundation of Greek astronomy, it is the Greeks to whom we must look for science:

> That the *Greeks* borrow'd the Foundation of their Astronomical Skill from the *Egyptians* and *Babylonians*, is a Point in which all their Writers are universally agreed, and need not be prov'd to so well acquainted with them as Yourself. This Concession of theirs, and the Want of understanding it with its proper and necessary Restrictions, has contributed amongst almost all Sorts of Writers to rob them of that Reputation they undoubtedly deserved. 'Tis to the happy Genius of that once glorious People, and that People alone, that we owe all that can properly be stil'd *Astronomy*; and 'tis but just to restore to them the Honour they have so long been depriv'd of. But in order to do this, we shall be at first oblig'd to step back into the remote and fabulous Ages of Antiquity, and for

---

[14]Nichols (1812), p. 535. Roger Long in his *Astronomy*, p. 648 described Heathcote's book as "an ingenious performance".

[15]For biographical details, see McConnell (2004) and the anonymous memoir in *The Gentleman's Magazine*, vol. 75 (1805), pp. 305–307.

one while as it were feel our Way in the Dark. Tedious and comfortless as this may be, we shall, however, as we advance, have the Pleasure to see the Morning of Science breaking in upon us at a Distance, and gradually increasing in Brightness, till at last it shines in the full Meridian Lustres, in which we now enjoy it.[16]

If Costard had little time for the claims of the Babylonians and Egyptians, his opinion of the Chinese as astronomers was even lower: "the Eastern Writers in general are much addicted to Fable and Romance".[17] As probably the only writer of the history of astronomy in the eighteenth century who could read Arabic, Costard was in a unique position to comment on Arabic astronomers. In a comparatively lengthy footnote stretching over pages 150–152 of his *A Letter Concerning the Rise and Progress of Astronomy*, Costard provided a concise overview of Arabic astronomy. Because his focus was on astronomy in antiquity, however, he did not go into greater depth. In 1777, Costard published a study of the passage in Ibn Yūnus, which describes observations of solar and lunar eclipses.[18]

Costard devoted a significant portion of his histories to the development of the constellations and the notion of the celestial sphere. His investigations included philological discussions of star names in Hebrew and Arabic. Although he celebrated the Greeks for first applying geometry to astronomy, Costard barely discussed lunar or planetary theory. Observations received more attention including analysis of some of the eclipse observations in the *Almagest* reproduced from Steete's *Astronomia Carolina*.

Interest in the history of science was strong in France also. In 1713 Pierre Rémond de Montfort wrote a letter to Nicolas Bernoulli in which he pointed out that it

> would be desirable if someone wanted to take the trouble to instruct us how and in what order the discoveries in mathematics have followed themselves, one after another; and to whom we should be obliged for them. The history of painting, of music, of medicine have been written. A good history of mathematics, especially of geometry, would be a much more interesting and useful work.[19]

The same sentiment could have been expressed about the history of astronomy. In 1693 Jean-Dominique Cassini had published *De l'Origine et du Progre's de l'Astronomie*, a short overview of astronomical history, but a detailed treatment of the subject in French did not appear until 1755 when Pierre Estève published *Histoire Generale et Particuliere de l'Astronomie*, a three-volume history of astronomy from its beginnings to the eighteenth century. Estève, a mathematician and astronomer, maker of faience, and member of the Société Royale des Sciences de Montpellier, was a prolific author of books on subjects ranging from architecture and fine art to mathematics, astronomical observations and music and language. In line with Estève's wide-ranging interests, the *Histoire Generale et Particuliere de l'Astronomie* was concerned not only with the origin and development of astronomy but also attempted to link the history of astronomy to other intellectual changes. History, for Estève, was an important subject, and the history of science was just as noteworthy as the history of wars and customs:

> One can never revoke doubts of the utility of history. In retracing the memorable actions of heroes, it inspires the noble ambitions of those that follow closely. The memory of victories gained and honours obtained, often awoke a love of glory and produced great men.
> Although from this first point of view the history of nations presents great advantages, one must not conclude it preferable to all other kinds of literature: because there is another aspect under which one should consider it, and which could make it very dangerous. The narration of crimes and villains may embolden a wicked person; seduced by the example of former models he may make his glory by surpassing them.
> The history of science is also instructive as the narration of battles and the portrayal of manners; but it does not expose the same dangers: the models which it presents are all worthy of being imitated. It is not disadvantageous

---

[16]Costard, *A Letter Concerning the Rise and Progress of Astronomy*, pp. 1–2.

[17]Costard, "A Letter … concerning the Chinese Chronology and Astronomy", p. 477.

[18]Costard, "Translation of a Passage in Ebn Younes".

[19]Cited by Peiffer (2002), p. 6, from whom I quote the translation.

to wish to pay attention to the sublime productions of Ptolemy, of Archimedes, of Copernicus and of Kepler; but Alexander and Charles XII are bad examples to follow for kings. Catilina, Marius, Sylla, and the other disturbers of the public repose, can corrupt ambitious citizens; it is more useful to ignore their names, than it is advantageous to know their actions.

After this reflection, one vows without doubt that the history of science is very much more useful than that of the revolutions of empires; but, one does not lack to object that it is not equally at the reach of all men: there will be something of truth in this objection, if by the history of science one does not intend a detailed narration of that which is more obscure in the knowledge of man; but this history, free of enigmatic language, presents with clarity the progress of the mind.[20]

Estève continued by explaining that he will present both a general history of astronomy, which will provide a "very clear" overview of astronomy focusing on astronomers and their observations, followed by a detailed history of astronomical ideas and theories.

Estève's claim of the importance of the history of science marks a significant departure from earlier histories of astronomy. For Flamsteed, the history of astronomy was of interest in providing information concerning earlier astronomical observations which might have contemporary uses and in justifying his own work at the Royal Observatory. Weidler's and Costard's interests were primarily antiquarian, as was Heathcote's, although he considered the history of astronomy and related philosophy a subject worthy of study in a university education. Estève, however, saw the history of science as a guide for the future. The great scientists of the past provided models to be aspired to. For Estève, science formed part of culture and *les progress de l'esprit humain* was the most important subject of history.[21] Estève's words echo those of other French thinkers of the time. In 1750, Turgot in his *Dissertation sur le progrès de l'espirit humian*, had claimed that only in mathematics (probably to be understood to include astronomy) could one see the sure and steady progress of the human mind. A similar view was presented by d'Alembert in the *Discours préliminaire* to the *Encyclopédie*.[22] For these French authors, the history of science was the history of scientific thought and acted as an index of the progress of intellectual thought.

Estève continued his preface by criticizing previous histories of astronomy, most notably Weidler's *Historia Astronomiae*, which he described as consisting of long passages copied from ancient authors who said anything relevant to the subject, without their being subject to critical reading. As a consequence, Estève complained that in the *Historia Astronomiae* Weidler "often gives within it fables and great absurdities, and reports with assurance matters of little likelihood".[23] Despite Estève's high ambitions and criticism of Weidler, his own history contained many mistakes ranging from straightforward historical errors such as his remark that the empire of the Caliphs was destroyed by the Mohammadeans,[24] to contemporary fables such as his story that Galileo had his eyes gouged out as part of his punishment by the Church.[25]

The "general" part of Estève's history dealt with many of the same topics as Costard and Weidler had discussed in their works: the myths of an extremely ancient astronomy, the pretended antiquity of Babylonian astronomy, the Greek philosophers, Ptolemy, and European astronomy down to the eighteenth century. The astronomy of the Arab lands was given only summary treatment and Chinese astronomy not mentioned at all. Etève's "particular" history was restricted to three main subjects: the stars and constellations, the theory of the earth, and solar theory. Lunar and planetary theories were not discussed in detail.

---

[20]Estève, *Histoire Generale et Particuliere de l'Astronomie*, pp. iii–v.

[21]Burke (1997), pp. 12–13 and 16–21.

[22]Laudan (1993), pp. 5–6.

[23]Estève, *Histoire Generale et Particuliere de l'Astronomie*, p. xi.

[24]Estève, *Histoire Generale et Particuliere de l'Astronomie*, p. 240. This error, along with several more, is pointed out by Montucla, *Histoire des Mathematiques*, pp. xxiii–xxiv.

[25]Estève, *Histoire Generale et Particuliere de l'Astronomie*, pp. 289–290; see Finocchiaro (2007), p. 113.

In 1758, 3 years after the publication of Estève's history, and a year after Lalande's study of the secular accelerations brought to an end for several decades work on establishing the size of the moon's secular acceleration from historical data, Jean Etienne Montucla published the first edition of his *Histoire des Mathematiques*.[26] Although ostensibly a history of mathematics, Montucla took as his subject the whole of the exact sciences. He illustrated this with a table of the division of mathematics into "pure" and "mixed" (i.e. applied) mathematics, the first of which was subdivided into arithmetic, geometry, and algebra, and the second into mechanics (statics and dynamics), astronomy (including spherical astronomy, astronomical geography, navigation, chronology, gnomonics, and theoretical astronomy), optics, acoustics, and pneumatics.[27] Like Estève, Montucla saw the history of science as an important part of the study of *l'esprit humain* and lamented that "our libraries are overburdened with verbose narrators of sieges, of battles, of revolutions".[28] His history was explicitly an examination of the progress of scientific thought; biography played only a minor role and was restricted to short notes about only the most important scientists, carefully placed in the text so as not to interrupt the central story of the progress of science.

The *Histoire des Mathematiques* was advertised by the publisher as early as 1754 but did not appear in print until 1758, too late to have any impact on Dunthorne or Mayer's use of historical eclipse observations. However, it is possible that Lalande, a life-long friend of Montucla's, may have seen or discussed parts of the unpublished work with Montucla before his own study of the secular acceleration was completed. Like Lalande, Montucla was educated at the Jesuit College of Lyon. He subsequently studied law in Toulouse before moving to Paris, where he gained a considerable reputation as a mathematician, scientist, and prolific author publishing books on the history of attempts to square the circle and the medical practice of variolation (inoculation by controlled infection with smallpox), as well as contributing to the *Gazette de France*.[29]

Montucla's *Histoire des Mathematiques* was a significantly more substantial work than any of the earlier eighteenth-century histories of astronomy not only in size (each of the two volumes contain more than 650 closely typeset pages) but also in breadth and depth of coverage. Eschewing Weidner's approach of citing everything he could find related to the history of astronomy, Montucla focused on the important stages in the development of science and discussed them in some detail. His discussion of Ptolemy's astronomy, for example, stretched over 20 pages, and whereas previous eighteenth-century histories of astronomy concentrated on the observations reported in the *Almagest*, Montucla also presented Ptolemy's cosmology and gave a creditable description of Ptolemy's models for the sun, moon, and planets, including discussion of the equivalence of epicycle and eccentric models for the sun and Ptolemy's introduction of the equant point in his planetary models. In the same fashion, Montucla devoted considerable attention to Copernicus and in his discussion of Tycho placed as much emphasis on Tycho's theoretical work and his objection to the heliocentric hypothesis as on his observations and instruments.

Several common characteristics can be identified in the histories of astronomy written by Flamsteed, Weidler, Heathcote, Costard, Estève, and Montucla. Despite the explicit statements of the importance and type of history favoured by the French authors, in content Estève's and, to a certain extent, Montucla's histories are not dissimilar to those of their English and German counterparts. Observation was seen as the part of astronomy whose history was worth recounting. Astronomical theory and, perhaps surprisingly, cosmology were, with the exception of Montucla, dealt with very superficially.

---

[26] An expanded second edition of the *Histoire des Mathematiques*, completed after Montucla's death by Lalande (with help from Lacroix and others), appeared in four volumes in 1799–1802.

[27] Montucla, *Histoire des Mathematiques*, pp. xxvi–xxviii; see also Swerdlow (1993), p. 302.

[28] Montucla, *Histoire des Mathematiques*, p. iii.

[29] For biographical details, see Sarton (1936) and Crépel and Coste (2005).

Indeed, it is striking how little these authors seem to have understood of, for example, the work of Ptolemy in contrast to astronomers of the seventeenth century. Longomontanus, Kepler, Bullialdus, and Riccioli all discussed Ptolemaic planetary models, epicycles, etc., in considerable detail, despite in Bullialdus's and Riccioli's cases their writing after Kepler's laws had become well accepted. Nevertheless, these authors were still deeply engaged with the Ptolemaic tradition. Within 50 years, however, that tradition had ended to be replaced by the new science of celestial dynamics founded on Newtonian theory. It would not be until the historical work of Delambre at the beginning of the nineteenth century that a detailed understanding of Ptolemaic astronomy would be reconstructed.

A second common thread in the writing of early eighteenth-century histories of astronomy was a reliance on the humanistic training of their authors, in particular their ability to use classical texts in Greek, Latin, and Hebrew, and their familiarity with classical and biblical literature. Perhaps surprisingly, these authors rarely applied their astronomical skills to the historical material, the notable exception being Flamsteed who examined some of the observations reported by Ptolemy and medieval Arabic authors in order to determine their accuracy. Weidler and Estève, both of whom made many observations of their own, did not attempt to analyse any observations.

Taking the early eighteenth-century histories of astronomy as a group, it is noticeable that in terms of content there is very little difference between the different histories. The authors all rely on the same basic sources of information about the history of astronomy and give the same account of the development of astronomy from its mythological beginnings, through the Babylonians and Egyptians to the Greeks, and up to the present day. The main differences concern the extent of the inclusion of Arabic and Chinese astronomy, though in all accounts these provide merely sidelines to the main story, and on the judgements passed on the relative merits of different astronomers and cultures.

Eighteenth-Century Views of Ancient and Medieval Astronomy

Babylonian Astronomy

For eighteenth-century Europeans, the name Babylon conjured images of a decadent city with a great tower reaching into the sky and luscious hanging gardens that were one of the seven wonders of the ancient world, ruled over either by the debauched queen Semiramis or the mad king Nebuchadnezzar, the land of the captivity of the Jews. The city itself had been lost to history. Beginning with Benjamin of Tudela in the twelfth century, European travellers to the near east had identified various sites as the city of Babylon. These sites included the ruins at Falluja near Baghdad, where remains of a ziggurat could be seen, and at Birs Nimrud to the southwest of Hillah (actually ancient Borsippa).[30] The true location of Babylon was, however, still known to the local Arabs who referred to the largest mound at the site as "Babil".[31] Probably armed with this knowledge, in 1616 Pietro della Valle visited the site and wrote the first detailed account of the ruins. Della Valle identified the site as Babylon and claimed that the mound was the remains of the biblical Tower of Babel and Herodotus's tower of the god Bel. Della Valle also wrote about the then undeciphered cuneiform inscriptions that he saw at Babylon. After della Valle's visit, the correct identification of the mounds at Hilla as the site of ancient Babylon gradually became accepted among European scholars.

For the general perception of ancient Babylon and its people, however, the discovery of the site of the city itself had negligible impact. Instead, the image of Babylon that was held by most people had

---

[30]Pallis (1956), pp. 43–44, Ooghe (2007), Reade (2008).

[31]Reade (1999).

been constructed from descriptions of the city and its inhabitants found in the Bible (more often than not seen through the lens of the first century A.D. Jewish historian Josephus).[32] Thus, the focus of European interest in ancient Babylon was on events such as the Jewish captivity, Daniel's prophecies, and the confusion of languages and the destruction of the Tower of Babel. These stories painted a picture of a decadent and corrupt society, an image reinforced by the use of Babylon as a metaphor for the sin, lust, and pride of imperial Rome in the Book of Revelation. References to Babylon in various classical sources seemed to confirm this view of Babylon as a city of depravity: Herodotus described a marriage market in Babylon where girls were lined up in order of beauty and sold, and accounts of a legendary queen Semiramis who had built Babylon gloried in telling of her sexual appetite, bedding soldiers from her army and having them killed the next morning, and of her dressing as a man to lead the army into battle—characteristics of a woman that would not generally have been admired (at least not publically) in eighteenth-century society.[33]

The early eighteenth century saw an upturn in interest in the near east. The publication in 1704 of a French translation of the "Thousand and One Nights" was followed by works of fiction by contemporary authors set in the ancient near east such as Voltaire's novels *Zadig* and *Semiramis*. These works presented a romantic vision of ancient Babylonia based upon Zoroastrian and classical legends. Handel's opera Belshazzar, first performed at the King's Theatre, Haymarket, London in 1745, mixed biblical and classical sources to tell a story about God's punishment of the sin of pride.[34] These examples of an increase in the use of Babylon for the purposes of story-telling, however, clearly illustrate that what was important about Babylon was not the history of the city itself, but the image that had been constructed of it. It mattered little where the historical Babylon was actually located—indeed, knowledge of the real Babylon would have taken away from the uses that people wanted to make of "Babylon", as it could never have lived up to the mystique that the city possessed.

The most common image of Babylon in European art and thought was of the Tower of Babel. According to the Book of Genesis, a great tower was built stretching up to heaven. When God saw the tower he scattered humanity throughout the world, confounding their speech so that all men no longer spoke the same language. The Jewish writer Josephus embellished the story, having God destroy the tower, and identifying Babylon as its location. The identification of the tower with Babylon was supported by Greek sources. Herodotus described a large, square tower at the centre of a temple in the middle of one half of the city.[35] Ctesius, in a fragment of his *Persica* found in Diodorus Siculus, claimed that Semiramis built a temple in the middle of the city to the Babylonian god Belus. Although he says that the building has fallen down, "It is agreed that it was extremely tall and that the Chaldaeans undertook observations of the stars in it since their rising and setting could be accurately monitored because of the building's height".[36] Perhaps surprisingly, this story is not discussed in accounts of Babylonian astronomy by eighteenth-century authors.

Artistic representations of the tower from the sixteenth to the eighteenth centuries range from iconic paintings by Bruegel, in which the gigantic tower emerges from Bruegel's native Flemish landscape, to the dark, foreboding vision of the tower dwarfing a scene of the desperate life of the poor ruled over with the sword by Frederick van Valkenborch, to attempts to depict an architecturally

---

[32] On the image of Babylon in European thought, see Lundquist (1995).

[33] On the portrayal of Semiramis throughout history, see Asher-Greve (2006).

[34] Lundquist (1995).

[35] Herodotus 1.181.

[36] Diodorus Siculus, 2.9.4; translation by Llewellyn-Jones and Robson (2010), p. 122. We have no evidence in support of Ctesias's claim that the Babylonians made observations from on top of the ziggurat; indeed, given what is known about the function of ziggurats, such a use is highly unlikely.

plausible tower constructed on the basis of Herodotus's account of the tower.[37] These paintings reinforced the image of Babylon known from the Bible and contributed to the general perception of the city and its people among Europeans, including the authors of histories of astronomy who almost uniformly have a low opinion of the Babylonians as a nation.

But what of the history of Babylonian astronomy itself? During the sixteenth and seventeenth century, writers such as Vergil, Cardano, Ramus and Savile had established a more or less standard genealogy of the origin and early history of the mathematical sciences.[38] According to Josephus, astronomical knowledge was first cultivated by Seth, the son of Adam, and his children in the days before the flood. When Adam predicted that the world would be destroyed once by fire and once by water, Seth and his sons made two pillars, one out of brick, the other stone, on which they described their knowledge of astronomy, one of which still remained in Josephus's day.[39] This knowledge then passed to either the Babylonians or the Egyptians and then to the rest of the world.

In the eighteenth century the story of Seth and the two pillars was given in the histories of astronomy written by Weidler and Heathcote, but other writers dismissed the story as a fable. Costard, for example, wrote:

> That Mankind began very early to lift up their Eyes to the Heavens, and observe that beautiful Canopy so richly adorn'd is not at all surprising; but that these Observations, before the Flood at least, contain'd any Thing more than meer Curiosity, may very easily be doubted. Josephus, fond of raising the Credit of his Nation, will needs make the immediate Descendants of *Seth* the original Authors of *Astronomy*. If he may be credited, they wrote too their Observations upon Pillars, one of Brick, and another of Stone, to preserve them secure against the Destruction, which *Adam*, it seems, had foretold them should, some time or other, put an End to all Things. The Extravagance and Inconsistency of this whole Account is such, as will justly excuse the saying any Thing farther upon it.[40]

Nevertheless, the age and origin of Babylonian astronomy was an issue of interest to Costard and others. Costard scoffed at the notion that the Babylonians had reports of observations stretching back 473,000 years (a figure taken from Diodorus Siculus II.31) was more willing to consider Simplicius's claim taken from Porphyry of Babylonian observations preserved for 1,903 years before the time of Alexander, but finally accepted Pliny's claim that Berossos says that there were observations for 480 years before his time. Since Berossos flourished about 56 years after the death of Alexander in 323 B.C., this puts the oldest Babylonian observations at about 800 B.C., not long before the oldest Babylonian eclipses reported in Ptolemy's *Almagest*. Although the Babylonians were famous in antiquity for their astronomy, Costard claimed that the Babylonians themselves "confess, that their Knowledge of the Heavens was brought to them from the *Egyptians*, by one Oannes or *Eubadnes* who *came out of* i.e. *up the* Euphrates".[41] This claim is a rather curious interpretation by Costard of fragments of the first book of Berossos's *Babyloniaca* in which knowledge is given to man by the man-fish Oannes. To my knowledge, Costard is unique in the history of interpretation of this fragment of Berossos in indentifying Oannes as an Egyptian.

Flamsteed posited an alternative origin for Babylonian astronomy. Noting the claim from Porphyry of Babylonian records of observations made 1,903 years before the time of Alexander, Flamsteed, however, reasoned that because Babylon does not appear in the Old Testament until the time of Isaiah, and (wrongly) that the early part of Ptolemy's king list gives only "Assyrian and Median" kings, the city was only founded just before Isaiah's time, and therefore

> if they had Observations made 1903 years before the taking of the town by Alexander, they had found and transcribed them from the Records of those Nations they conquered, about or after this time. [Indeed], 'tis most

---

[37] Minkowski (1991), Wegener (1995), Albrecht (1999), Seymour (2008).

[38] See, for example, Grafton (1997), Popper (2006), Goulding (2010).

[39] Josephus, *The Antiquities of the Jews*, 2.69–2.71.

[40] Costard, *The Rise and Progress of Astronomy*, pp. 2–4.

[41] Costard, *The Rise and Progress of Astronomy*, p. 50.

probable from the Jews, whose Solemn Festivals being tied to certain Days of the Moon, might occasion them to be more attentive to her Motions, and probably to keep Records of Eclipses than other Nations, who had no Festivals that like them, were tied to the full Moons, and therefore had less occasion to take notice of them.[42]

This is part of Flamsteed's wider claim that the Babylonians learnt all their astronomy from other nations, either the Medes or the Jews. For example, the earliest eclipse records from Babylon given in the *Almagest* date from 720 B.C., "*that is a year after the first Captivity of the Jews* at the hands of Tiglathpileser" and which

happening so near the Captivity, seems to intimate that the Chaldeans received their astronomy from the Subjugated Jews, who having a knowledge of Dialling cannot be thought ignorant of the courses of the Sun and Moon. Probably, the Chaldeans learned from them their Saron [Saros], or the return of lunar eclipses after 223 Lunations, for the Jewish State, 'tho[ugh] now declining, had continued about 1,000 years from Moses. During that time the Jews might have easily learned this term of eclipses, from their frequent observations of them at the Full moons and imparted it to their new masters and conquerors, to gain more respect from them.[43]

For Flamsteed, the Babylonians were not the renowned astronomers they were often said to be in antiquity but were merely the recipients of the astronomical knowledge of others. In Flamsteed's view, even the most famous Babylonian achievement in astronomy, the discovery of the Saros cycle of 223 synodic months, often simply called the "Chaldean period" by his contemporaries, had been learnt by the Babylonians from the Jews.

Eighteenth-century writers had five sources of information about Babylonian astronomy to draw upon: (1) the Bible and early Jewish works, in particular Josephus; (2) classical writers, in particular Herodotus, Diodorus Siculus, and Pliny; (3) the writings of Greek astronomers, especially Geminos, who discusses Babylonian lunar theories, and Ptolemy, who reports a number of observations made in Babylon; (4) the fragments of Berossos; and (5) histories of astronomy written during the fifteenth, sixteenth, and seventeenth centuries. None of these sources presented a clear picture of Babylonian astronomy and they often contradicted one another. For example, the model of lunar motion attributed to Berossos by Cleomedes and Vitruvius, a physical model in which the moon is ball of half-fire moving on a rectangular path, has no connection with the aspects of Babylonian lunar theory discussed by Geminos. The challenge facing eighteenth-century writers of histories of astronomy, therefore, was to try to assemble a coherent story of Babylonian astronomy from this incomplete collection of misfitting pieces.

As I have discussed, the story Flamsteed constructed reduces Babylon to the recipient of a now lost Jewish astronomy. Flamsteed relied almost exclusively on Jewish sources for his account of Babylonian astronomy; the only other material that he makes use of are the Babylonian eclipse reports given in Ptolemy's *Almagest*. Costard, by contrast, drew on the classical historians, the Bible, and the *Almagest* for his history of Babylonian astronomy in his *Rise and Progress of Astronomy*. Costard allowed that the Babylonians were the originators of their own astronomy, but in his view they were merely

Diligent observers of the Heavens, no Doubt, they were, and carefully mark'd every Phenomenon that could come to their Knowledge ... But one sees the wide Difference between this, and a Science based upon strict demonstrative Principles.[44]

And even though diligent, the observations of the Babylonians were in Costard's view still very crude. Their eclipse records, for example, were only "bare *Registers* of what had been observed",[45] and they had no knowledge of how to predict eclipses: "That the *Theory* of the *Moon's Motions* was at all

---

[42]Chapman and Johnson (1982), p. 30.

[43]Chapman and Johnson (1982), pp. 31–32.

[44]Costard, *The Rise and Progress of Astronomy*, pp. 22–23.

[45]Costard, *The Rise and Progress of Astronomy*, pp. 21–22.

known so early as this, or that the *Chaldeans* were ever capable of *calculating* or *predicting* an *Eclipse*, is more than can be made appear from any good Authority now extant",[46] somewhat at odds with his explanation a few pages later of Thales prediction of an eclipse by means of the "Chaldean Saros"[47] (perhaps we are meant to assume that the Saros is a Greek discovery, but was attributed by later authors to the Chaldeans?). He also dismissed the claim found in Diodorus Siculus that the Babylonians could predict the appearances of comets as unfounded. Costard's portrayal of Babylonian astronomy as purely observational, and not particularly accurate at that, was a necessary part of his larger goal of demonstrating the originality and genius of Greek astronomy, an aim he is quite explicit about at the beginning of his work.

Costard returned to the history of Babylonian astronomy in *A Further Account of the Rise and Progress of Astronomy amongst the Ancients, in Three Letters to Martin Folkes, Esq* published in 1748. The three letters, numbered II, III, and IV following on from the letter which contained *The Rise and Progress of Astronomy*, all largely concern Babylonian astronomy: letter II discusses Babylonian lunar theory and eclipses, letter III the origin of the constellations, and letter IV the astronomy of myths. Letter II is of particular concern to us here. Costard presented a very different picture of Babylonian astronomy to that he had given in *The Rise and Progress of Astronomy* 2 years earlier. Relying now largely on Geminos and Ptolemy, Costard discussed the Babylonian theory of the moon, something he had denied even existed in his earlier account of Babylonian astronomy. Without acknowledging his change in position, Costard now placed Babylonian lunar theory at the beginning of the western tradition:

> Having already spoke of the *Chaldeans* as observing *Eclipses*, which seems to imply their having some knowledge of the *Moon's* Motion, it may not be unuseful or unentertaining, perhaps, to consider distinctly what that Knowledge was, and in what manner acquired. By this means we shall see how far the Theory of that Luminary is indebted to them, and from what slender Beginnings it has grown up to the Accuracy to which it is at present arrived.[48]

Costard next described these "slender Beginnings" of the lunar theory. The Babylonians, he now explained, discovered lunar anomaly and measured the length of the anomalistic month at 27 days 13 hours and 20 minutes, understood that eclipses of the moon were caused by the moon entering the Earth's shadow and that eclipses of the sun were caused by the interposition of the moon, discovered the Saros and knew it to be 6,585 1/3 days in length, measured the length of the synodic and draconitic ("periodical") months, and modelled the variable velocity of the moon. This list of Babylonian discoveries in lunar theory includes almost everything that Costard also mentioned in his discussion of Greek lunar theory. It is a remarkable about-face by Costard. The Babylonians have gone from not having a lunar theory to having one that is to all intents and purposes is the same as that of the Greeks who Costard lauds as the originators of scientific astronomy. How did this happen? Costard obtained most of his information on Babylonian lunar theory from Geminos, but it is clear that he already knew Geminos when he wrote his earlier history: Costard cited Geminos several times in other contexts in that earlier work. It seems impossible not to conclude that Costard deliberately downplayed the achievements of Babylonian astronomy in *The Rise and Progress of Astronomy* in order to support his aim of restoring the image of Greeks as the first true scientists. I can offer no explanation why, 2 years later, he provided this corrective account of Babylonian astronomy. Costard's description of Babylonian astronomy in letter II of his *Further Account of the Rise and Progress of Astronomy* is the most detailed, reasonable, and accurate account of Babylonian astronomy published in the eighteenth century.

---

[46]Costard, *The Rise and Progress of Astronomy*, p. 22.

[47]Costard, *The Rise and Progress of Astronomy*, p. 85.

[48]Costard, *A Further Account of the Rise and Progress of Astronomy*, p. 3.

In contrast to Costard's detailed discussion of Babylonian astronomy in *A Further Account of the Rise and Progress of Astronomy*, most other accounts of the time are either short or lack any attempt at analysis. Weidler devoted 22 pages to the Babylonians and the Phoenicians in his *Historia Astronomiae*, but these mainly consist of quotations from any classical source which mentions the "Chaldeans". Heathcote condensed Weidler's material into a single long paragraph, stretching over about a page and a half of his *Historia Astronomiæ*, in which he dismissed the great antiquity of Babylonian astronomy and noted the association of the "Chaldœos" with astrology in Greek and Roman sources. Estève used several pages to dismiss the supposed great antiquity of Babylonian astronomy before presenting a very critical account of Babylonian astronomy, which he took to be purely astrological, referring to the astronomers as "Des Mages de la Chaldée". Estève draws mainly on the Bible and references to Berossos in classical sources, but frequently becomes confused about the contradictions found in these sources. For example, if Berossos was responsible for bringing Babylonian astronomy to the Greeks and was said to have lived around the time of Alexander, why does Herodotus say that the Greeks learnt of the pole and the gnomon from the Babylonians. Estève concluded that there must have been two men named Berossos. Montucla is brief but somewhat fairer in his discussion of Babylonian astronomy, dismissing the claims of the great age of Babylonian astronomy, but accurately describing the Babylonian observations reported by Ptolemy and the discussion of Babylonian lunar periods in Geminos.

In general, eighteenth-century accounts of Babylonian astronomy focused on the question of its origin and antiquity and very general remarks about the transmission of Babylonian astronomy to the Greeks found in Herodotus and other classical historians. The eclipse observations from Babylon reported in Ptolemy's *Almagest* were often mentioned only in the context of establishing the date of the beginning of Babylonian astronomy. Any discussion of the nature and accuracy of these accounts was usually reserved for a later chapter on Ptolemy. Only Costard, in his *Further Account of the Rise and Progress of Astronomy*, and, briefly, Montucla discussed the evidence for Babylonian theoretical astronomy from Geminos. The reason for the absence of discussion of Babylonian theoretical astronomy from these histories may well have been because the authors sought to present a history of the progression of astronomy from its early, empirical beginning, through the Greek miracle of scientific astronomy, down to the culmination of astronomy in contemporary times as represented by Newton. There was no place in this story for Babylonian theoretical astronomy, either chronologically (postdating the beginnings of Greek astronomy, and simultaneous astronomical traditions are not part of a story of progress), or thematically (the Babylonian astronomy described by Geminos being not obviously geometrical, an assumed prerequisite of scientific astronomy). The image of Babylonian astronomy constructed by these histories was one of unreliability (its pretended antiquity), superstition (its intimate connection with astrology), and crudeness (inaccurate observations and lack of geometrical theories).

Ancient Greek Astronomy

Whether the origins of astronomy lay in Babylonia, Egypt, or with the Jews, for all eighteenth-century authors who wrote about the history of astronomy, the *science* of astronomy began with the Greeks. For example, Costard, in the introduction to his history of astronomy which I have quoted above, made it clear that, despite the claims made by classical authors, the debt of Greek astronomy to the Babylonians and Egyptians was minimal. Only the pole, gnomon, and the division of the day may have come from Babylon, as Herodotus had reported, although Costard wished that Herodotus had been clearer in his discussion of the matter. For the rest, however, Costard made the somewhat unusual argument that "Because *Babylon* lying so far within Land, and out of the Way of Correspondence with the *Greeks*, to whom we owe our knowledge of antiquity, we cannot expect that they should have borrow'd much from thence".[49] This claim would seem to be contradicted by the presence of Babylonian observations in Ptolemy's *Almagest*, which Costard was fully aware of, but fits in with Costard's wider narrative where the Greeks are firmly positioned at the centre of ancient astronomical knowledge.

Costard and others placed the beginning of scientific astronomy with Thales and Pythagoras. Thales, "the next Person that we hear of, as treating upon *Astronomical* subjects, and from whose Time, indeed, we may properly date all that truly deserves that Name", Costard wrote, had visited Egypt "for Improvement, which would incline one to think, that it was but about this Time, that that Country began to be famous for *Science*".[50] In other words, only when a Greek visited Egypt did the Egyptians start to do science. Thales was well known for predicting an eclipse of the sun that stopped a battle between the Lydians and the Medes. Interestingly, Costard again contradicted his claim that the Greeks learnt nothing from the Babylonians by suggesting that Thales made his prediction using the Babylonian Saros cycle:

> To predict an *Eclipse*, of the *Sun* especially, is a Work of Labour and Difficulty, and required better Tables than, it is to be feared, *Thales* was furnished with. When I say *better Tables*, it is only on the Supposition that he had any at all: For, as seems to be most probable, and in which I find others likewise concur with me, he rather collected it only by attending to the *Chaldean Saros*; a Period consisting of 223 *Lunations*, after which Time the *Eclipses* of the *Sun* and *Moon* return in the same Order again.[51]

Flamsteed made the same claim that Thales had learnt the Saros from the Babylonians. He also noted that Pythagoras travelled widely in Egypt and the near east learning "their languages, three sorts of letters and their Sacred Rites, along with their Astronomy and Geometry".[52] However, even if the Greeks had learnt astronomy and geometry from the Egyptians or the Babylonians, the application of geometry to astronomy was a purely Greek innovation. Discussing the motion of the planets, Costard wrote:

> From this Time, however, it is probable their Motions began to be observed, and *Geometry* to be applied to the Purposes of *Astronomy*: A Thing, as far as appears, unattempted by the *Egyptians* and *Babylonians*; and yet without it could never be reduced to a Science.[53]

Thus for Costard, astronomy is only a science when it is subject to geometrical analysis. Although not always presented explicitly, this equivalence between astronomy as a science and the application of geometry to astronomical theory is assumed by all other eighteenth-century writers of histories of astronomy.[54]

The culmination of the geometrical approach to astronomy in the Greek period came with the work of Hipparchus in the second century B.C. and Ptolemy in the second century A.D. Both figures are celebrated for their achievements in all eighteenth-century histories of astronomy, although the space accorded to their work is generally much less than might have been expected. Indeed, in most eighteenth-century histories less detail is given about Hipparchus and Ptolemy than about the writings of men such as Plato, Archimedes, and Pliny. This is almost certainly due to the mathematical nature of Hipparchus' and Ptolemy's works; the authors of eighteenth-century histories may have seen the geometrical approach of Greek astronomers as the very definition of scientific astronomy, but few wanted to get involved with it themselves.

---

[49]Costard, *The Rise and Progress of Astronomy*, pp. 51–52.

[50]Costard, *The Rise and Progress of Astronomy*, pp. 88–89.

[51]Costard, *The Rise and Progress of Astronomy*, pp. 94–95.

[52]Chapman and Johnson (1982), p. 34.

[53]Costard, *The Rise and Progress of Astronomy*, pp. 129–130.

[54]Lamentably, this naïve assumption persists in many ill-informed histories of astronomy today, despite the efforts of historians working on non-Greek traditions of astronomy.

Ptolemy's writings provided the main source of information about both his own astronomy and the astronomy of his predecessors (especially Hipparchus, but also Timocharis and other Greek observers). The late medieval recovery of the *Almagest* in Europe resulted in the printing of two Latin translations and a Greek edition of the text by the middle of the sixteenth century: in 1515 Petrus Lichtenstein published Gerard of Cremona's twelfth century translation under the title *Almagesti Cl. Ptolemei Pheludiensis Alexandrini, astronomorum principis, opus ingens ac nobile, omnes cœlorum motus continens*; this was followed in 1528 by the Giunti printing house's publication of George of Trebizond's translation of 1451 with the title *Almagestum seu magnae constructionis mathematicae opus plane divinum Latina donatum lingua ab Georgio Trepezuntio*; finally in 1538 a Greek edition prepared by Simon Gryneus and Joachim Camerarius was published in Basel with the title *Claduii Ptolemaei Magnae Constructionis, id est Perfectae coelestium motuum pertractionis, Libra XIII*.[55] Appended to the Greek edition were the preserved books of Theon of Alexandria's commentary on the *Almagest*.

In addition to the *Almagest* several printed editions or translations of other works by Ptolemy were published in the sixteenth and seventeenth century. The Greek text of the *Tetrabiblos*, Ptolemy's treatise on astrology, was edited by Camerarius and published in Nuremburg in 1535, 3 years before the Greek edition of the *Almagest*. Despite debates over its authenticity, some scholars being unwilling to believe that a work of astrology was written by the same man who was capable of writing the *Almagest*, the *Tetrabiblos* remained a popular text and was sometimes taught alongside the *Almagest* in Renaissance universities.[56] The *Tetrabiblos* was the first of Ptolemy's works to be published in the vernacular, an English translation by John Whalley appearing in 1701 with the title *Ptolemy's Quadupartite; or, Four Books Concerning the Influences of the Stars*. The *Geography* was translated into Latin by Jacopo d'Angelo in 1406. Many printed versions appeared from 1475 on. The Greek text was edited by Erasmus and printed in 1533 in Basel.

During the fifteenth and sixteenth centuries, Ptolemy came to epitomize ancient astronomy. Little was known about the man himself; often he was confused with the ruling Ptolemies of Egypt enabling him to be seen both literally and figuratively as the "king of astronomers". Images of Ptolemy appear regularly in the frontispieces to astronomical books of the period, frequently showing him wearing a royal crown and holding some astronomical instrument. Depictions of Ptolemy were not confined to scientific books, however. He appeared in paintings by artists including Raphael and in sculptures at Ulm and Florence cathedrals.[57] Indeed, such was the common association of Ptolemy with astronomy that his name became attached to a widely read work called *The Compost of Ptholomeus Prynce of Astronomye* which first appeared in 1532 and was republished many times during the sixteenth century. The *Compost* contained the pirated text of the astronomical sections of an earlier work called the *Kalender of Shepherdes* put together by the English publisher Robert Wyer. Whenever the word "Shepherd" appeared in the text, Wyer simply substituted "Ptholomeus" or "Astrologian".[58] This work was a very low-level introduction to astronomy and astrology, bearing no link with Ptolemy's astronomy. Detailed understanding of the true Ptolemy's astronomy was widespread among astronomers of this period, however, largely thanks Regiomontanus's *Epitome of the Almagest*, written in 1462 and printed in 1496. Ptolemy continued to feature prominently in the frontispieces to astronomical books until the late seventeenth century. He is among the astronomers standing within the temple to knowledge in the frontispiece of Kepler's *Rudophine Tables* published in 1627 (Fig. 5.1), appears as a rather forlorn, defeated figure in the frontispiece of Riccioli's *Almagestum Novum* of 1651 (Fig. 5.2), and is debating astronomy with Tycho and Copernicus in Johann Gabriel Doppelmayer's *Atlas Coelestis* of 1742.

---

[55] Pedersen (1974), p. 21.

[56] Rutkin (2010).

[57] Derome (2000–2001), Fuchs (2009).

[58] Johnson (1937), p. 73.

Fig. 5.1 Frontispiece to Kepler's *Rudolphone Tables*. Ptolemy is sitting to the far right (Courtesy John Hay Library, Brown University Library)

By the beginning of the seventeenth century Ptolemy's astronomical models had become obsolete but the observations recorded in the *Almagest* were still useful for studying precession and changes in the obliquity of the ecliptic. However, close study of these observations was tarnishing Ptolemy's reputation as the greatest astronomer of antiquity. The Danish astronomer Christian Longomontanus, formerly Tycho's assistant on Hven,[59] was highly critical of the quality of Ptolemy's observations and raised the suspicion that several of Ptolemy's "observations" were actually the results of calculation.[60]

---

[59] For biographical details, see Christianson (2000), pp. 313–319.

[60] Swerdlow (2010) provides a detailed study of Longomontanus's analysis and use of the solar observations in the *Almagest*.

Fig. 5.2 Frontispiece to Riccioli's *Almagestum Novum*. Ptolemy's is lounging forlornly watching the balancing of the Copernican and the Tychonic world systems (Courtesy John Hay Library, Brown University Library)

In his *Astronomia Danica* of 1622, Longomontanus claimed that Ptolemy adjusted his observations of solstices and equinoxes in order to obtain agreement with Hipparchus's determination of the length of the tropical year. In support of this claim, Longomontanus says that Ptolemy's observation of the parallax of the moon was:

> Half a degree and more above the true parallax, for no other reason (as I believe) than that he pass off upon posterity as genuine that hypothesis of the moon he previously established himself or, if you prefer, received from his predecessors, and only once confirmed by his computation. But now, I ask, what will be the prohibition

150

Eclipfes XIX veterum Aftronomorum Babyloniorum, Hipparchi, & Ptolomæi: quas refert ipfe Ptolomæus lib. 4. μαθηματικ. τωνζάξεως. Ad Meridianum Vraniburgi reductæ, & Tempori medio accommodatæ.

Anni Impp. Regg. Ma-giftr. & Period.	An, Na-bon	Menfes	Temp. Medium D. H I	Locus ☉ verus S. g ° ' "	Anomalia ☽ g ° ' "	Locus ☾ computatus S. g ° ' "	Diff côp. ab obfer. ° ' "	Motus la-tit. ☽ à ☊ g ° ' "	Digiti obferua-ti. D. '	Digiti comp. D. '	Latitudo ☽ ° ' . "
Mardo kempadi. A. I	27	Thoth.	29. 7. 26	♓ 21. 53. 39	67. 15. 28	♍ 21. 48. 45	m 4. 54	1 51. 40	Total.	19. 10	7. 52. B
Mardokempadi. A. 2	28	Thoth.	18. 9. 59	♓ 11. 6. 54	13. 39. 0	♍ 11. 2. 36	p 0. 42	9. 10. 46	D. 3. A	3. 10	49. 15. B
	28	Phamenoth.	15. 6. 23	♍ 1. 17. 57	164. 11. 34	♓ 0. 58. 13	m 19. 24.	183. 48. 35	D 6. p B	6. 17	45. 40. M
Nabopolaffati An. 5	117	Athyr.	27. 15. 34	♈ 14. 49. 15	341. 19. 54	♎ 14. 56 27	p 6. 11	170. 37. 41	D. 3. A	3. 20	48. 35. B
Cambyfis An. 7	225	Phamenoth.	17. 8. 52	♋ 16. 53. 35	29. 16. 20	♍ 16. 26. 45	m 26. 50	351. 9. 57	D. 6. B	5. 15.	45. 47. M
Darij Hyftafpis. F. A. 20	246	Epiphi	28. 9. 35	♍ 22. 17. 31	4. 14. 12	♉ 21. 59. 19	m 18. 11	170. 33. 12	D. 3. A	3. 3	48. 57. B
Darij Hyftafpis F. A. 31	257	Tybi	3. 9. 13	♈ 8. 55. 24	101. 37. 18	♎ 29. 15. 19	p 19. 56	169. 54. 49	D. 2. A	2. 40	53. 34. B
Archonte Phanoftrato.	356	Thoth.	26. 17. 9	♋ 17. 14. 7	229. 0. 54	♊ 27. 12. 41	m 1. 25	349. 15. 22	β exχ η.	2. 10	55. 49. M
Archonte Phanoftrato.	366	Phamenoth.	24. 6. 56	♉ 10. 53. 6	29. 1. 40	♋ 10. 46. 51	m 6. 15	171. 13. 46	. 640	40. 28. B
Archonte Euandro.	367	Thoth.	16. 8. 21	♋ 16. 24. 46	182. 17. 40	♊ 16. 17. 54	m 6. 52	356. 54. 4	Totalis.	19. 42	16. 10. M
Periodi. II. Calipp. A. 54	547	Mefori	16. 5. 36	♍ 26. 12. 57	105. 37. 17	♓ 26. 2. 48	m 10. 9.	173. 31. 34	Dig. 7. A	9. 8	33. 36. B
Period. II. Calipp. A. 55	548	Mechir	9. 12. 5	♓ 25. 47. 29	110. 43. 1	♍ 26. 6. 36	p 19. 7	5. 1. 51	Totalis	16. 14	25. 56. B
Period. II. Calipp. A. 55	548	Mefori	5. 11. 55	♍ 25. 20. 9	250. 36. 30	♓ 15. 10. 4	m 10. 5	181. 24. 35	Totalis.	19. 45	6. 55. M
Philometoris Anno 7	574	Phamenoth.	27. 11. 51	♉ 6. 4. 1	164. 56. 29	♍ 6. 22. 27	p 18. 26	138. 23. 14	D. 7. B	7. 13	43. 24. M
Period. III. Calipp. A. 37	607	Tybi	2. 9. 2	♒ 4. 51. 15	180. 5. 34	♌ 5. 9 50	p 18. 59	10. 41. 4	D. 3. A	2. 57	55. 15. B
Adriani An. 8	872	Pachon	17. 7. 2	♈ 14. 33. 40	251. 15. 4	♎ 14. 19. 25	m 14. 15	169. 17. 23	D. 2. A	2. 12	55. 25. B
	17	880	Payni	10. 9. 46	♉ 14. 31. 17	315. 36. 3	♍ 14 31. 27	p 2. 10	315. 34. 27	Total.	11. 0	23. 45. A
Adriani	19	882	Choeac	2. 9. 28	♎ 16. 16. 41	66. 0. 29	♈ 16. 24. 36	m 2. 6	185. 31. 56	D. 9. B	11. 6	29. 4. A
	20	883	Pharmuthi	19. 14. 48	♓ 14. 57. 23	147. 31. 57	♍ 15. 32. 29	p 35. 6	351. 29. 32	D. 6. B	6. 27	44. 57. A

Fig. 5.3 Bullialdus's analysis of the eclipse observations in Ptolemy's *Almagest* (Bullialdus, *Astronomia Philolaica*, p. 150) (Courtesy John Hay Library, Brown University Library)

from suspecting that here he was of the same intention, and relied upon those equinoctial observations of the sun which served his purpose, but the others, of which it is very likely he made many more, he entirely concealed.[61]

Longomontanus's claim was that Ptolemy selectively chose observations which agreed with his theory, even adjusting them where necessary to produce the desired result.

On pp. 50–51 of the second part of the *Astronomia Danica* Longomontanus analysed three groups of three eclipses used by Ptolemy in book IV of the *Almagest* to investigate the moon's anomaly: three seen in Babylon in 721 and 720 B.C., three observed in Alexandria in 201 and 200 B.C. and which had been used by Hipparchus (Longomontanus incorrectly states that these were observed by Hipparchus), and three observed by Ptolemy himself in A.D. 133, 134, and 136. Longomontanus lamented that the accounts of the eclipses in the *Almagest* were vague; sometimes when a time for the eclipse was given it was unclear which phase of the eclipse it related to. He says that the errors he finds in the time of the eclipses may have been caused by the transmission of the account of the eclipse from the ancient authors, by changes in the motion of the sun and moon, or from inaccurate waterclocks. Furthermore, Longomontanus says, the errors in the times of the eclipses observed by Ptolemy are much greater than those of the other eclipses.

A similar criticism of Ptolemy's reports of eclipse observations was made by Bullialdus in his *Astronomia Philolaica* of 1645. In book 3, Chapter 7, for each of the 19 eclipses recorded in the *Almagest* Bullialdus compared the magnitude and the moon's longitude at mid-eclipse calculated from the local time of the eclipse adjusted to the meridian of Uraniborg with his calculations of the circumstances of the eclipses (Fig. 5.3; for a summary of his analysis, see Table 4.1 above).[62] Bullialdus found that although some of the earliest eclipses agreed quite well with his calculations, the difference

---

[61] Longomontanus, *Astronomia Danica*, II, p. 33. The translation quoted here is taken from Swerdlow (2010), p. 176.

[62] See Chap. 4 above for Struyck's summary of Bullialdus's analysis.

between the moon's longitude deduced from the observed time of the eclipse and his calculated longitudes was in several cases almost 20′ and for the very latest eclipse, observed by Ptolemy himself in A.D. 136, was in excess of 35′. Bullialdus then turned to six of the eclipses in more detail. In *Almagest* IV.11 Ptolemy discussed two sets of three eclipses used by Hipparchus to determine the size of the lunar anomaly. For the first pair, Hipparchus had found that the time interval between the midpoints of the first two eclipses was 177 days and 13 3/5 equinoctial hours and between the second and third eclipses 177 days and 1 2/3 hours. For the second pair, Hipparchus had found 178 days and 6 equinoctial hour and 176 days and 1 1/3 equinoctial hours. Ptolemy noted that these intervals led to a different value of the anomaly from that which he had found, and also to two different values depending upon whether one used the eccentric or the epicyclic hypothesis, which should not be the case. He argued that Hipparchus's deduction of the time of the eclipses from the observation reports was incorrect and that the actual time intervals between these eclipses was 177 days and 13 3/5 equinoctial hours, 177 days and 2 equinoctial hours, 178 days and 6 5/6 equinoctial hours, and 176 days and 2/5 equinoctial hours; Hipparchus made errors in these time intervals of up to almost an hour. The differences between Hipparchus and Ptolemy's interpretations of these eclipse reports made Bullialdus suspicious of Ptolemy, especially since Hipparchus's interpretations agreed better with Bullialdus's own calculations. He concludes:

> For there is no certainty in these ancient observations of within a third of an hour: and I fear that Ptolemy falsely signaled all of the times, as he himself professes that he has changed some of the other observed intervals.[63]

Bullialdus was similarly critical of Ptolemy's accounts of the observations of solstices and equinoxes used to determine the length of the year and of the star positions used to determine the rate of precession.

Such suspicions concerning the observations reported by Ptolemy are largely absent from the histories of astronomy written in the eighteenth century. Flamsteed asserted that Ptolemy's star catalogue was simply that of Hipparchus with a correction of 2° 40′ to account for the 265 years between their two epochs at his rate of precession of 1° per century, but that "Nowhere, however, does he say that he has actually observed all the stars in this catalogue, or that he himself has arrived at their positions from his own observations",[64] despite the fact that Ptolemy does say precisely this in *Almagest* VII.4. Flamsteed also noted the errors in some of the solstice and equinox observations but attempted to explain them as due to atmospheric refraction or the design and construction of the instruments used to make the observations. Costard referred to some of the observations in the *Almagest*, but made no remarks on their reliability, and Weidler and Heathcote did not mention the observations at all. Estève also did not discuss Ptolemy's observation reports, but he did devote 14 pages to Ptolemy and his work in astronomy and geography. For Estève, Ptolemy represented the highpoint of ancient science, and the *Almagest* the most important book of ancient astronomy:

> Astronomy then languished for some time on all sides of the Mediterranean, when at the beginning of the second century of the Christian era, Ptolemy, who kept the school of Alexandria, wrote a work which had the title *la grande Composition*, and that the Arabs have named the *Almagest*. This astronomer enlightened through precise geometry, comparing his observations with those of Hipparchus, of Timocharis, those which were known in Greece & those of the Babylonians. The consequences of this comparison are tables which determine the movements of the stars, the planets & the sun. ... The *Almagest* can be regarded as a collection of almost everything that the ancients have seen in the heavens; & if we did not possess this book, all of that the ancients did in the positive science of the stars would have been lost to us. One must know how to appreciate the author of the work of which we speak, for not having composed in the manner of the preceding Greeks. ... Ptolemy was one of those rare geniuses whose sublime designs are of advantage to society.[65]

---

[63] Bullialdus, *Astronomia Philolaica*, p. 152.

[64] Chapman and Johnson (1982), p. 46.

[65] Estève, *Histoire Generale et Particuliere de l'Astronomie*, pp. 194–196.

This is a strong endorsement from Estève, despite him clearly not having studied Ptolemy's astronomy in detail. He was only able to refer to the epicycle and eccentric models in general terms and did not mention Ptolemy's equant model which had so troubled Copernicus and Kepler, or give any quantitative information on any of Ptolemy's astronomy.

In general, for eighteenth-century authors Ptolemy was to be praised as the greatest astronomer of antiquity but his work itself was little read and less understood, in striking contrast to the mid-seventeenth century where, if no longer taken seriously as an alternative to the astronomical systems of Copernicus, Tycho, and Kepler, it was still fully understood. Within a mere 50 years, Ptolemy's astronomy changed from being part of a living tradition of astronomy, fully connected to contemporary astronomical thinking, to a piece of history, celebrated as the first mathematical astronomy but its details forgotten. The only exception to this general lack of interest in or understanding of the details of Ptolemy's astronomical theories is to be found in Montucla's *Histoire des Mathématiques*. Montucla devoted 24 pages to Ptolemy, covering Ptolemy's astronomy as well as his work on optics and harmonics. Montucla's account provides a fairly clear and accurate account of Ptolemy's solar, lunar, and planetary theories, including a description of Ptolemy's use of combined eccentric-epicycle models and the equant point. Montucla also discussed the instruments Ptolemy described in the *Almagest*. Perhaps surprisingly, however, he had little to say about the observational records given by Ptolemy.

After Ptolemy, we are told in eighteenth-century histories, astronomy went into decline. Of later Greek writers on astronomy, only Theon of Alexandria generally gets mentioned. Theon wrote a commentary on the *Almagest*, but was best known as the father of Hypatia. The story of Hypatia—her life as the "first" female mathematician and her death at the hands of a mob—caught the imagination of sixteenth- and seventeenth-century historians, often obscuring the history of Theon's contributions of astronomy.[66]

Medieval Arabic Astronomy

European knowledge of Arabic astronomy was founded largely on the few texts by medieval Arabic authors that had been translated into Latin.[67] Most important among these were the introduction to astronomy by al-Farghānī (Alfraganus), several translations of which were published in Latin, Thābit ibn Qurra's work on trepidation, again published in several translations with the title *De Motu Octave Sphere*, and an influential work by al-Battānī (Albategnius) usually known in Europe as *De Scientia Stellarum*. Regiomontanus, Copernicus, and other astronomers of the Renaissance drew heavily on these works both for their understanding of Ptolemy and for the observations and ideas they contained.

The seventeenth century saw an increase in interest in "Arabick" learning among Europeans scholars.[68] Knowledge of Arabic was deemed useful for understanding related Semitic languages, which had application in the study of the Hebrew Bible, for the recovery of ancient texts that had been translated into Arabic and lost in the original Greek, and for the promotion of European national interests in the near east. Among the scholars with an interest in Arabic were scientists such as John Greaves and Edward Bernard (both of whom held the Savilian Chair of Astronomy at Oxford). These early scientific Arabists discovered important geographical and astronomical data in Arabic manuscripts.

---

[66] On the role played by Hypatia in histories of science, see Goulding (2010).
[67] For a survey of Latin translations of Arabic astronomical works, see Carmody (1956).
[68] Russell (1994), Toomer (1996).

Considerable numbers of Arabic manuscripts were brought to Europe during the seventeenth century. Of particular importance was the collection brought back to Leiden by Jacobus Golius in 1629. A catalogue of these manuscripts published in 1630 revealed a number of astronomical and mathematical texts among which were several Arabic translations of Greek works that were partly or fully lost in the original, including Ptolemy's *Planetary Hypotheses*, and Apollonius' *Conics*,[69] and a copy of the *zīj* of Ibn Yūnus. Numerous requests were made by astronomers for access to these manuscripts, or for Golius to send details of any observations they contained, but Golius by and large kept the manuscripts to himself. Only Wilhelm Schickard was eventually allowed access to the Ibn Yūnus manuscript; Schickard's Latin translations of reports of three eclipses recorded in that manuscript were eventually published by Curtius in his *Historia Coelestis*.

Following Golius's death, the Oxford scholar Edward Bernard travelled to Leiden to buy Arabic manuscripts from his estate. From these manuscripts, Bernard was able to gather many astronomical parameters, observations and star positions, which he shared with other astronomers. Thomas Streete, the author of the *Astronomia Carolina* wrote to Bernard asking for any "Ancient observations Astronomical, the older the better, but since the time of Ptolemy, or not published (and wrought) by him". Any observations of planetary positions, lunar eclipses, conjunctions of the moon with a star, or star catalogues, Streete continued, "are very much to be desired, for without them there can never be any certaine limitation of the Middle Motions or true places of all the Planets and Starrs until some hundreds of yeares after our time".[70] Unfortunately, the eclipses which Streete sought in order to refine the lunar theory were not to be found among the manuscripts Bernard had collected.

By the end of the seventeenth century, interest in Arabic had waned and eighteenth-century accounts of Arabic astronomy generally relied purely on those works that have been published in Latin translation. Historical accounts usually began with the translation of the *Almagest* into Arabic in the ninth century. Before this time, according to Estève, the Arabs were too occupied with warfare; only when they had successfully conquered the near east did they have time to devote to the sciences. Under the patronage of the Caliph al-Ma'mūn, astronomers such as Ḥabash and al-Farghānī were said to have produced astronomical tables and books, but these were purely based in the work of Ptolemy. All Arabic astronomers achieved, according, for example, to Estève or Costard, was to correct some of the errors in Ptolemy's astronomy. In particular, they correctly revised the rate of precession from 1° per century to 1° per 66 years. Astronomy may have been "studiously cultivated"[71] by the Arabs, but they were not presented as having made significant progress in its development.

The one area where the Arabs were seen as having contributed to astronomy was in observation. Costard wrote approvingly of the accuracy of Arabic observations. Estève noted that many Arabic observations are known, although he felt that they had been put to little use:

> The Arabs have several observations of celestial phenomena, but generally there was little that came out that improved the theory of astronomy. In all time there may be Princes who protect the sciences, but they are not able to create geniuses capable of making useful progress. It is not that nature cannot produce in one century what it has produced in another, I say only that great men are extremely rare in any field whatsoever.[72]

Flamsteed included a fairly extensive discussion of Arabic determinations of the obliquity of the ecliptic and observations of the positions of stars in his history of astronomy. In addition to the observations found in those works which had been translated into Latin, Flamsteed used a collection of star positions found in Arabic texts made by Bernard which was included in Bernard's letters to Dr Robert Huntingdon and which Huntingdon had communicated to the Royal Society. Flamsteed was

---

[69] Toomer (1996), pp. 48–49.

[70] Bodley MS Smith 45, p. 35; quoted by Mercier (1994), pp. 190–191.

[71] Costard, *The Rise and Progress of Astronomy*, p. 150.

[72] Estève, *Histoire Generale et Particuliere de l'Astronomie*, pp. 224–225.

disappointed to find that several Arabic star catalogues were dependent either upon Ptolemy's catalogue or upon each other. For example, describing Nasīr al-Dīn al-Tūsī's star positions, he wrote:

> As for Nassir Oddin's Longitudes of his Stars, their differences agree so well with Ptolemy's that if he did not borrow them from him, he seems to have had a great regard to him in stating them, and for their Latitudes too. Tho' he seems to be a diligent Observer by his determination of the Sun's greatest Declination, yet in some of them he is as erroneous as Ptolemy, and not so well as Abolchusan [sic] that preceded him and Olegh Beigh that followed him.[73]

Flamsteed's interest, however, was restricted to observations of star positions and the obliquity; he did not discuss other types of observations made by Arabic astronomers (for example the well known eclipse observations reported by al-Battānī), and had no interest in the theoretical tradition:

> Thebit Ebn Corrah, Arzachel, Alfraganus and Alphonsus King of Arragon made Astronomical Tables for representing the motions of the Sun and Planets betwixt the years
> Hegira 278 and 678
> A.D. 900 1300
> in which some of them followed Ptolemy's previous determinations, others Albatanus, and some endeavoured to answer both by strange Contrivances but none of them having given us any good Observations, I have nothing to say to them.[74]

In general, the great interest that had been shown in Arabic manuscripts by seventeenth-century scientists had waned during the eighteenth century, and along with this decrease in interest came a decrease in the perceived contribution of Arabic astronomers to the development of astronomy. The Arab astronomers were at best viewed as "diligent observers", though often not even that. They were seen as unable to break away from the Ptolemaic tradition, and as such their contributions to astronomical theory were of little value. The real only interest that Arabic sources might hold was if they preserved observations that might be useful because of their age. In particular, Curtius's publication of Schickard's translation of three eclipse records in Golius' manuscript of Ibn Yūnus still raised the prospect of more observations being found in that text. In an echo of Streete's letter to Bernard, Euler wrote to Capsar Wetstein concerning this manuscript (first assuming it was held in Oxford, before correctly giving its location as Leiden). Wetstein sent the letter for publication in the *Philosophical Transactions*:

> Monsieur *le Monnier* writes to me, that there is, at Leyden, an Arabic Manuscript of *Ibn Jounis* (if I am not mistaken in the Name, for it is not distinctly wrote in the Letter), which contains a History of Astronomical Observations. M. *le Monnier* says, That he insisted strongly on publishing a good translation of that Book. And as such a Work would contribute much to the Improvement of Astronomy, I should be glad to see it publish'd. I am very impatient to see such a Work which contains Observations, that are not so old as those recorded by *Ptolemy*.[75]

It would not be until the end of the nineteenth century, however, before the remaining observations recorded in the manuscript were published.

European Astronomy

The European Renaissance within astronomy was generally assumed by eighteenth-century authors to have begun with Regiomontanus, Peurbach, and Walther and to have culminated with Copernicus's heliocentric model, Tycho's observations and Kepler's discovery of elliptical orbits. Eighteenth-

---

[73]Chapman and Johnson (1982), p. 54.

[74]Chapman and Johnson (1982), p. 54.

[75]Euler, "Concerning the Gradual Approach of the Earth to the Sun", p. 203.

century histories of astronomy focused almost exclusively upon these individuals, even to the extent of often ignoring Galileo, something that would be unthinkable in histories of astronomy written today. Only Estève and Heathcote devoted significant space in their histories to Galileo's discoveries with the telescope.

The authors of eighteenth-century histories of astronomy were often hampered when discussing pre-Renaissance astronomy by the scarcity of original sources. Only select works were preserved and in some cases, for example ancient Babylonian astronomy, historians had to rely for information on second-hand accounts preserved in non-native sources. Writing about Renaissance astronomy, however, most of the primary source material—the books written by the astronomers concerned—were readily available. For example, a printed edition of Regiomontanus's *Epitome of the Almagest* was published in 1496, three editions of Copernicus's *De Revolutionibus* had been published by the middle of the seventeenth century, and books by Tycho and Kepler were widely available. In addition, collections of observations made by Renaissance astronomers had been extracted from manuscripts and published. For example, observations by Regiomontanus and Bernard Walther were available in a collection of short astronomical and mathematical treatises written by Georg Peurbach, Regiomontanus, and Walther that was edited and published by Johannes Schöner with the title *Scripta Clarissimi Mathematici M. Ioannis Regiomontani* in Nuremberg in 1544. Schöner, a keen astronomical observer, professor of Mathematics in Nuremberg and correspondent of Rheticus and Copernicus, had bought or otherwise acquired many of Regiomontanus and Walther's manuscripts, and during the 1530s and 1540s edited several of these for publication. Schöner's edition contained several typographical errors, which were pointed out to other astronomers. Nevertheless, the observations it contained, especially Walther's solar data, were used extensively by later astronomers including Tycho, Mästlin, Kepler, Flamsteed, and Lecaille.[76] Schöner's edition of Regiomontanus and Walther's astronomical observations were republished either in whole or in part (and sometimes with further typographical errors) in W. Snel's *Coeli et siderum in eo errantium Hassiacae*, Curtius's *Historia coelestis* and Riccioli's *Almagestum novum* and *Astronomia reformata*.

Tycho's observations were known from both his own publications and later collections. Tycho included data from a small number of observations in his *Astronomiae Instauratae Progymnasmatum* published in three parts in Prague in 1602 (the year after Tycho's death). After the well-known dispute over ownership of Tycho's observational logbooks, Kepler used many of Tycho's observations in constructing his planetary models. A substantial number of observations from these logbooks were collected and published by Curtius in *Historia Coelestis* of 1666, with a few more published by Lalande in his 1757 paper on the secular accelerations.

For eighteenth-century historians of astronomy, Regiomontanus marked the beginning of the Renaissance. Overlapping in time with the latest Arabic astronomers, Regiomontanus's was seen as the first astronomer to move beyond Ptolemy's astronomy through his skills in mathematics and his realization of the importance of accurate and extensive observational data. Flamsteed acknowledged Regiomontanus's work only briefly:

> In that Century lived Regiomontanus, his scholar [pupil] Bernard Walther, and Copernicus, each of which promoted Astronomy beyond what any before them had done. … Regiomontanus's Observations commence in the year 1457 and were continued 'til 1474, and he Died in 1476, having done more to the Promotion of this Science than any that lived in that Age before, by his Skill in Geometry and Mechanicks.[77]

Flamsteed discussed Walther's observations in more detail, noting that he "seems to have been very diligent in his Observations … and they are of good use, 'tho they could not be so exact as those taken by Tycho and Hevelius with their more convenient sextants".[78] Flamsteed studied Walther's observations

---

[76] Kremer (1981).

[77] Chapman and Johnson (1982), pp. 54–55.

[78] Chapman and Johnson (1982), p. 55.

of the solar meridian zenith distance finding differences with Flamsteed's own solar tables of less than 4'.

Flamsteed turned next to Copernicus, about whom he has little to say other than to remark that his work led to Reinhold's Prutenick Tables, which were a great improvement on the medieval Alphonsine Tables:

> Copernicus's Works had not long been extant but the sincere Lovers of Truth embraced them, and Erasmus Reinhold published Tables of the Coelestial Motions agreable to his Hypothesis, which supported the Motion of the Earth. Now began the Combat betwixt the two Hypotheses, the Alphonsine Tables were found to deviate enormously from the Heavens and the Prutenick of Reinhold's commonly much less, but very remarkably, the Almighty Providence that will not suffer his rationall Servants to be Ignorant of Useful Truths, and that Mankind might be fully convinced of his Wisdom and the folly of Human interventions in Mans contriving to represent the Motions of the Heavens otherwayes than he had formed them.[79]

Copernicus, however, serves but as a prologue for the entry of the hero into Flamsteed's story: Tycho Brahe:

> The All wise Archetect, I say, having raised up Copernicus to revive the true System of the Heavens, within less than a Century after him, sent Tycho Brahe, a Noble Dane into the world with a Spirit fitted for this purpose.[80]

For Flamsteed, Tycho was sent by God to discover the true structure of heavens through observation (conveniently forgetting Tyhco's objection to the motion of the Earth). And it was the improvement in observational techniques that Flamsteed saw as the real marker of scientific progress. Flamsteed ended this paragraph praising God's wisdom as manifested in the birth of Tycho, quoting a psalm singing of God's creation.

Flamsteed devoted the whole of the third chapter of his history to Tycho and his observations. He reprinted extensive passages from Tycho's descriptions of his instruments in the *Astronomiae Instauratae Mechanica*, discussed Tycho's account of his discoveries in the same work, and commented on the observations collected by Curtius in the *Historia Coelestis*, concluding that "our noble Tycho Brahe, Prince of Astronomers of his time, who shall be revered for ever with the greatest honour by all heaven- and earth-born creatures".[81] For Flamsteed, Tycho represented the ideal of what an astronomer should be: dedicated in observation, innovative in the design of instruments, and assembling a body of empirical data that could provide the essential foundations for the development of astronomical theory. The parallels with Flamsteed's own career in astronomy were clear for his readers to see: an individual labouring on the thankless task of gathering observational data, working with royal patronage but having to spend his own money to further the project, while other astronomers (Newton and Halley in Flamsteed's case, perhaps Kepler in Tycho's) waited like vultures to pick over the empirical data before it was complete.

A somewhat different history of European astronomy to Flamsteed's story of progress in observation was given by Estève. For Estève, Copernicus not Tycho represented the watershed between ancient and modern science. It is therefore appropriate, he explained, to end the first book of his general history with the decline of Arabic science and begin the second with Copernicus. Estève placed great importance on Copernicus's development of the heliocentric model, seeing it as both simpler than the astronomy of Ptolemy (which is not necessarily true), and an act of bravery: "Copernicus was the first astronomer who combated the prejudices of long-standing tradition".[82] Nevertheless, Estève devoted almost as many words to Tycho as to Copernicus, praising him for the quality of his observations and his scientific method:

> Although ancient astronomers were occupied in observing the stars, we can say that Tycho far surpassed them, because no one before him provided resulting observations as accurate nor also as continuous as that found in his

---

[79]Chapman and Johnson (1982), p. 59.

[80]Chapman and Johnson (1982), p. 59.

[81]Chapman and Johnson (1982), p. 94.

[82]Estève, *Histoire Generale et Particuliere de l'Astronomie*, p. 251.

works. At times astronomers have observed to justify an unlikely system, but not Tycho Brahe who measured the motions of the heavenly bodies in order to ensure only the facts. He had a kind of disinterestedness to any particular opinion, which persuades us of the truth of his observations.[83]

The image of Tycho as the disinterested observer, collecting unbiased observational data, rather than observing just to establish the parameters of an existing theory, was for Estève an important stage in the development of a progressive science.

Unlike Flamsteed, Estève discussed not only Tycho's observations but also his rejection of the Copernican system and Tycho's own cosmological model in which the Earth remained stationary at the centre of the universe, orbited by the sun which carried with it the five planets. Estève was sympathetic towards Tycho for developing this system, saying that at the time there was little evidence to choose between to the two. Nevertheless it was on the quality of Tycho's observations that his legacy rested.

Overall, the history of Renaissance European astronomy presented by eighteenth-century authors focused on two issues: the development of a systematic programme of accurate celestial observation by Tycho and Copernicus' proposal for a sun-centred universe. Both were seen to have led to the creation of modern astronomy through the work of Kepler. But the balance between the presentation of Tycho's observations and Copernicus' cosmology differed between authors. Flamsteed focused almost exclusively on the observational tradition, praising Walther and especially Tycho for their contributions, whereas Estève and other more philosophically inclined authors such as Heathcote gave as much if not more attention to Copernicus's cosmology as to Tycho's observations. What is common to all, however, is the absence of any discussion of the technicalities of the astronomical theories developed by Copernicus or Tycho.

Chinese Astronomy

Jesuit missions to China from the late sixteenth to the eighteenth century provided a conduit for the exchange of scientific knowledge between China and Europe. The supposed superiority of western astronomical methods was used by the Jesuits to attempt to gain influence in the Chinese court, which relied upon astronomical calculation for the timing of ritual activities, the promulgation of a calendar and the regulation of the heavens, all important activities for the maintenance of a dynasty's mandate to rule. Selected western scientific texts were translated into Chinese and western instruments, including the telescope, were brought to China.[84] The Jesuits also wrote about China for a European audience, providing information about the country's history, culture, geography, philosophy, and science.[85]

The accounts of Chinese astronomy found in early Jesuit works usually focused on the mythological origins of the study of astronomy (for example, the story of Hi and Ho who failed to predict an eclipse), the names and identification of the constellations, and ancient astronomical observations. Less interest was shown in contemporary or recent observational or mathematical astronomy, which only a couple of centuries earlier had been far ahead of European astronomy. The Jesuits, or the native Chinese who were instructing them, instead gathered ancient astronomical observations from sources such as the *Chunqiu* 春秋 ("The Spring and Autumn Annals", often attributed to Confucius) or the *Shujing* 書經 ("Classic of History"). These works are historical annals rather than scientific texts. The

---

[83] Estève, *Histoire Generale et Particuliere de l'Astronomie*, p. 264.

[84] D'Elia (1960).

[85] Hsia (2009).

Chunqiu is a historical chronicle of the state of Lu covering the period from 722 B.C. to 481 B.C. It contains reports of 36 solar eclipses in amongst historical accounts of state.[86] However, these reports are not recorded as straight-forward observations but in memorials and discussions of portents. The *Shujing* is a history of China from mythological times down to the early first millennium B.C. It contains various stories of extremely ancient astronomical events, including an eclipse of the sun supposedly dating to the third millennium B.C.

A significant improvement in European knowledge of Chinese astronomy came through the work of the Jesuit Antoine Gaubil (1689–1759), who lived in Beijing from 1722 until his death. Gaubil collected a far wider range of information on Chinese astronomy than previous Jesuits. His research encompassed all aspects of Chinese astronomy including units of time, the names and positions of constellations, calendrical astronomy and the assumed lengths of the year and month, and records of observations, including the large number of ancient and medieval records of astronomical phenomena found in the dynastic histories. Throughout his time in China, Gaubil maintained an active correspondence with European scholars. For example, as I will discuss in chapter 7, Gaubil wrote to Joseph Delisle about the date of the ancient eclipse reported in the *Shujing*; Gaubil's letters on this subject were forwarded to various scholars including Tobias Mayer.

Many of Gaubil's researches on Chinese astronomy (along with his own observations made in China) were collected and published by Souciet in the three volume *Observations Mathématiques, Astronomiques, Géorgraphiques, Chronologiques, et Physiques; Tirées des Anciens Livres Chinois, ou faites nouvellement aux Indes, à la Chine & aileurs, par les Pères de la Compagnie des Jesus* published between 1729 and 1732.[87] Gaubil's research on Chinese astronomy also provided the main source for the chapter on astronomy in Jean-Baptiste Du Halde's popular *Description Géographique, Historique, Chronologique, Politique, et Physique de l'Empire de la Chine et de la Tartarie Chinoise* of 1736, which was published in an English translation by R. Brookes in the same year with the title *The General History of China*. Du Halde's presentation of Chinese astronomy included a discussion of the ancient eclipse in the *Shujing* dated by Gaubil to 2155 B.C., the eclipses in the *Chunqiu*, which Du Halde praised as "Of the thirty-six Eclipses of the Sun, related by *Confucius*, there are but two false and two doubtful; all the rest are certain",[88] astronomical instruments, the production of the calendar and associated rituals, and the constellations. Du Halde, however, did not discuss Gaubil's collection of ancient and medieval observations from the dynastic histories.

Eighteenth-century historians of astronomy dealt with Chinese astronomy in a number of different ways. Flamsteed and Heathcote simply ignored it. Estève, more puzzlingly, discussed the invention of the compass by the Chinese before it was known in Europe, but had nothing to say about Chinese astronomy itself. But Weidler and Montucla each devoted chapters of respectable length to China. Both drew exclusively on Jesuit works, especially Gaubil, and provided a fair summary of what was known. Montucla argued that the long history of Chinese science, as presented by the Jesuit historians, was not evidence for its greatness, but indicated instead that the Chinese had made little progress in the pursuit of science. If they had done otherwise, then science at the time of the Jesuits' arrival should have been more advanced than European science, since it had had longer to develop. The answer, Montucla said, was that the occurrence of individual geniuses who furthered the development of science was a particularly Greek and European phenomenon.

A more brutal assault on the history of Chinese astronomy was made by George Costard in a letter published in the *Philosophical Transactions* for 1747.[89] Costard had not included China in his

---

[86] For an English translation and study of the eclipse reports in the *Chunqiu*, see Stephenson and Yau (1992).

[87] On the writing of *Observations Mathématiques, Astronomiques, Géorgraphiques, Chronologiques, et Physiques* and Gaubil's relations with Souciet, see Hsia (2009), pp. 121–128.

[88] Du Halde, *The General History of China*, v. 3, p. 80.

[89] Costard, "Concerning the Chinese Chronology and Astronomy".

earlier treatise on the history of ancient astronomy, in which he dismissed Babylonian and Egyptian contributions to astronomy. Costard felt that all the "Eastern Writers", that is, the Babylonians, the Egyptians, and the Chinese, "in general are much addicted to Fable and Romance".[90] The history of Chinese astronomy has been written from the accounts of the Jesuits, but Costard felt that the Jesuits had been too credulous in believing what they had been told by the Chinese:

> The best Accounts we have received of *China* are owing to the *Jesuits*. But those Accounts themselves are, I am afraid, to be frequently received with great Caution. These Fathers have been sometimes, perhaps, not suffi-ciently versed in *European* or *Chinese* Learning, or both, to give us proper Information. At other times, it may be, they have been too much prejudiced in Favour of their Converts, or had Ends to serve, of which the World hath not been properly enough apprised. To have propagated their Religion only in a barbarous and uncultivated Nation, would not have been so much for the Credit of the Mission, as to have been able to introduce it among a People civilized and polished by Arts and Literature.[91]

This was a strong attack not only on Chinese astronomy but on the Jesuits as well. Costard accused the Jesuits of being either incompetent scientists and historians who had been taken in by their Chinese informants, or, worse, of having deliberately given a false picture of Chinese history in order to make their own work as missionaries seem more important and difficult that it actually was.

At the heart of Costard's criticism of Chinese astronomy was a basic scepticism concerning the truthfulness of Chinese observational reports. Costard gave several examples to support this view, but the most damning came from contemporary times:

> And to put it out of all Doubt, that the *Chinese* are capable of obtruding upon the World *fictitious* Observations, we need no other Authority still than that of the *Learned Fathers* themselves. In the Year 1725 the *Missionaries* sent into *Europe* an Account of an Approximation of the four Planets *Jupiter*, *Mars*, *Venus*, and *Mercury*. Such planetary Conjunctions, it seems, in *China* are look'd upon as happy Omens of good Fortune to the Prince upon the Throne. The *Chinese* therefore, as if brought up at the Court of *Versailles*, with a true *French* politeness, in Compliment to their Sovereign, mark'd in the Registers a Conjunction of all the 7. This false Account of an imaginary Conjunction, as the *Learned Jesuit*, himself observes, may, in future Times, be the Occasion of very great Errors.—To the *Chinese*, I suppose he means; for in *Europe* the Danger will be but small; where there are better Tables, exacter Accounts, and more accurate Observations, than the most sanguine *Jesuit* will pretend to be among the *Chinese*. But if they would venture at recording such a spurious Observation, at a time when they were sure of being detected, what may we not suspect them to have been guilty of, when they had none to con-front them; and how little may we presume they know of the Uses to be made of *Celestial* Observations?[92]

Finally, Costard argued that because the Chinese treat eclipses and other astronomical events as omens, they must be "very *bungling Astronomers*"[93] indeed, and that "I think we need but little more to convince us of the small Acquaintance of the *Chinese* with *Astronomy*".[94]

The history of Chinese astronomy as it was written in the eighteenth century was constructed from a highly selective search in Jesuit accounts of Chinese astronomical texts for references to observation. As I have discussed above, with the exception of Gaubil, the Jesuits were largely uninterested in the Chinese tradition of *li* 曆 (calendrical or mathematical) astronomy, and their search for astronomical observations was focused upon looking for extremely ancient records. And on the whole, except again for Gaubil, the Jesuits did not assemble large numbers observations from the *tianwen* 天文 (observa-tional astrology) or the *li* 曆 (calendar) treatises in the dynastic histories. Instead, they concentrated on searching through annals and other historical works. As a result, the quality, extent, and reliability of the records they found were justifiably questioned by late seventeenth- and eighteenth-century European

---

[90]Costard, "Concerning the Chinese Chronology and Astronomy", p. 477.

[91]Costard, "Concerning the Chinese Chronology and Astronomy", pp. 477–478.

[92]Costard, "Concerning the Chinese Chronology and Astronomy", pp. 481–482.

[93]Costard, "Concerning the Chinese Chronology and Astronomy", p. 483.

[94]Costard, "Concerning the Chinese Chronology and Astronomy", p. 486.

astronomers. The historians of science were, with the exception of Montucla, mainly interested in the development of observational astronomy as a marker of scientific progress, and so were unimpressed with what they found in the works such as Du Halde's *Description Géographique, Historique, Chronologique, Politique, et Physique de l'Empire de la Chine et de la Tartarie Chinoise*, which referred only to the ancient and unreliable observations and did not discuss Gaubil's collection of medieval observations.

When investigating the secular acceleration of the moon, Dunthorne, Mayer, and Lalande decided not to use Chinese reports of eclipses. In part this was simply because the eclipse records that were widely known—those from the *Shujing* and the *Chunqiu*—were not well suited to studying this problem as they lacked timings of the eclipse contacts. In principle the simple fact than an eclipse at a given location puts constraints on the magnitude of the acceleration (a technique Dunthorne used for one of the lunar eclipses reported by Ptolemy), but concerns about the reliability of the records seem to have overcome their potential usefulness. Mayer did investigate the account of the very ancient eclipse from the *Shujing*, but argued that the account was too problematical to be of use. The Chinese records were therefore ignored in attempts to determine the moon's secular acceleration, dogged by their reputation for being unreliable, imprecise, and even forged. Ironically was visible, if scholars had considered the extensive set of medieval observations collected by Gaubil and published in the third volume of *Observations Mathématiques, Astronomiques, Géorgraphiques, Chronologiques, et Physiques* they would have found many timed observations of lunar eclipses from the fifth to the thirteenth century—exactly the time period missing from the European and Arabic corpus that astronomers were frequently lamenting. But Gaubil's collection of these records seems to have been relatively unknown at the time. It would not be until the late twentieth century that the importance of these observations was recognized and they were applied to the problem.

Chapter 6
The First Detailed Study of the Moon's Secular Acceleration: Richard Dunthorne

I cannot conclude without acknowledging my obligations to Mr. Dunthorne. To those who have looked deep into these subjects it is enough to have given his name. Others may please to be informed that there is a person at Cambridge who, without the benefit of an Academical education, is arrived at such a perfection in many branches of learning, and particularly in Astronomy, as would do honour to the proudest Professor in any University: and yet, notwithstanding this supreme skill in a science so difficult, and so important, Humanity and Modesty, the most engaging diffidence of himself, and readiness to advance others, are parts of his character not only more excellent and amiable, but more peculiar and distinguishing.

—William Ludlum, *Astronomical Observations made in
St. John's College, Cambridge* (Cambridge, 1769), preface.

By the late 1740s Halley's claim that there existed a secular acceleration of the moon had been accepted by scientists including Newton, Whiston, and Euler, but the existence of the acceleration still needed to be demonstrated and its size determined from empirical data. The first person to undertake a detailed examination of historical records in order to answer these questions was Richard Dunthorne. Dunthorne developed innovative new techniques for using ancient eclipse observations that allowed him to place constraints on the size of the moon's acceleration. Through his careful study of historical records, Dunthorne proved without question the existence of the acceleration and determined that the coefficient of the secular equation is about $10''$ per century^2, close to the true value.

Richard Dunthorne

Richard Dunthorne was born in 1711 in Ramsey in Huntingdonshire where his father worked as a gardener.[1] On the completion of his education at the free grammar school in Ramsay, Dunthorne moved to the nearby town of Alconbury to run a private school. In his spare time, Dunthorne cultivated an interest in mathematics and astronomy. He corresponded several times with *The Ladies Diary*, a yearly periodical founded in 1704 by John Tipper and which from 1709 onwards included

---

[1] Brief, and at times inaccurate, biographical accounts of Dunthorne may be found in Cooper (1863), Lynn (1905), Skempton (2002), Clerke and McConnell (2004), and Croarken (2007).

J.M. Steele, *Ancient Astronomical Observations and the Study of the Moon's Motion (1691–1757)*,
Sources and Studies in the History of Mathematics and Physical Sciences,
DOI 10.1007/978-1-4614-2149-8_6, © Springer Science+Business Media, LLC 2012

"Arithmetical Questions (which) are as entertaining and delightful as any Subject whatever".[2] Dunthorne, along with many accomplished mathematicians of the day,[3] submitted several answers which were published between 1735 and 1742 and set prize questions in 1738 and 1739. Dunthorne also sent to the *Diary* his calculations of the circumstances of a transit of Mercury over the face of the Sun on 31 October 1736 made using Streete's *Astronomia Carolina*, and was one of several people, including Charles Leadbetter, whose calculations of an upcoming eclipse were also given in the 1736 *Diary*.[4]

By the middle of the 1730s Dunthorne had started constructing a set of lunar tables based upon Newton's lunar theory as presented in Gregory's *The Elements of Astronomy, Physical and Geometrical*. He soon learned, however, that Samuel Rouse, a draper in Market Harborough, was also attempting to construct Newtonian lunar tables. Rouse was an enthusiastic astronomer and mechanic who would later develop the bent-lever balance, but whose drapery business eventually went bankrupt.[5] Dunthorne and Rouse began corresponding, forming a friendship that would last for some thirty years.[6] Rouse sent Dunthorne his tables in order that they might be compared. Dunthorne's reply is dated 13 July 1737 (Fig. 6.1):

Sir
I have compared our Tables together, and find they agree very well, except that of the moon's Variation in the Octants, in which there is a difference of 8 seconds when most, and that I have the 6th equation of $\mathbb{C}$ $2'25''$ when greatest, and have also Contrary Titles in my Table of the 3° Equation of $\mathbb{C}$; in the other Tables the difference is rarely more than 1 or 2 seconds, only in the Tables of the Equation of the $\mathbb{C}^s$ Centre where there is sometimes a difference of 3 or 4 seconds, which however is not very considerable. I have likewise compared your method of finding the Inclination of the Lunar Orbit to the Ecliptic with my Table, and find the difference not above 7 or 8 seconds when Greatest. I have from the Physical Theory enquired into the Annual Equation of the Moon's Horizontal Parallax, and also into that Equation of the Horizontal Parallax which depends on the aspect of the Sun and Lunar Apogee (of which we were speaking) and find 'em both very inconsiderable. U purpose (God willing) to enquire from the Physical Theory, into the quantities of some other Equations not hitherto taken notice of by Astronomers, particularly into the Annual Equation of the Inclination of the Lunar Orbit to the Ecliptic, all which I will take the first oppertunity of communicating to you. I am ready to think the Titles add & subtract in the 7th Equation should be contrary to those set down by Dr. Gregory supposing it to be those Inequalities explained by Dr. Gregory in the 16th. Prop of his 4th. Book. I desire your thoughts on that matter. I have sent you the first part of my Table of the Equation of Time, and will take the first oppertunity of communicating whatsoever else I promised, but am in haste by reason of my not timely notice of this oppertunity; I am with humble service to your self and Mrs. Rouse (only wishing we were nearer, that we might have more frequent Correspondence)

Your most humble Serv[t]
Rich[d]. Dunthorne

Alconbury
13 of July 1737

Pray direct for me to be left with Mr. Whitechurch
At Mr. Hardings in Suffwick[7]

---

[2] *The Ladies Diary* 1709, p. 25.

[3] Albree and Brown (2008).

[4] *The Diary Miscellany*, pp. 33, 51, 53–54, 64, 74, 86, 91, 93, 99, 141, 149.

[5] *The Gentleman's Magazine*, 1823, vol. 93, p. 89.

[6] Rouse had his correspondence with Dunthorne and other astronomers bound together in a slim volume now held at the Museum for the History of Science, Oxford (MS Museum 95). The volume begins with a handwritten copy of William Ludlum's tribute to Dunthorne given in the preface to his *Astronomical Observations Made in St. John's College* (quoted at the head of this chapter). More than 30 letters sent by Dunthorne to Rouse are listed in Rouse's handwritten contents page, mostly dating to between 1737 and 1753, but the last from 1764. Unfortunately, the last few letters have been torn out of the binding. The left edges of the pages of Dunthorne's letter of 29 April 1964 concerning ancient Greek calendars (bound as pp. 177ff) are all that is preserved of that letter, and the whole of the later material is lost. It is not known when this damage occurred.

[7] MS Museum 95, p. 31.

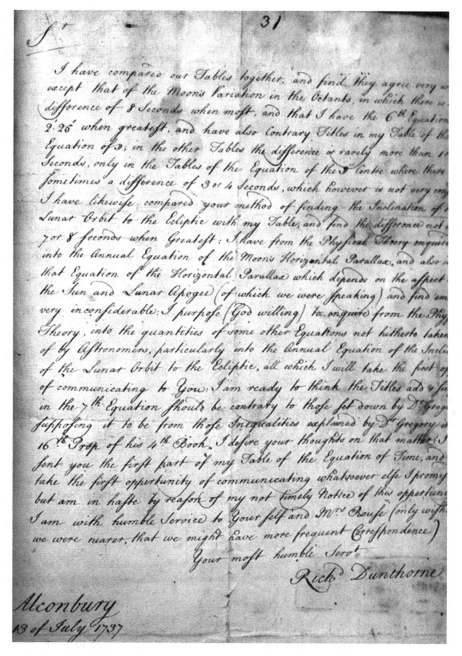

Fig. 6.1 Letter from Richard Dunthorne to Samuel Rouse dated 13 July 1737 (MS Museum 95, p. 31) (Courtesy Museum of the History of Science, Oxford)

This first letter sent by Dunthorne to Rouse illustrates many of the common features of their correspondence. Both Rouse and Dunthorne frequently ask the other to check their work, to look for mistakes in the theory, discrepancies with observation, and mathematical slips. Both are very happy to share their results. In later letters Dunthorne passed on collections of observational data, laboriously copied out by hand (with his letter of 10 February 1737/1738, Dunthorne enclosed eight tightly packed pages of solar observations made by Flamsteed and Walther). Finally, Rouse and, especially, Dunthorne feel starved of the chance to discuss their work. Dunthorne ends many of his letters with a statement that he hopes Rouse will be able to visit him soon, that he will be able to travel to Market Harborough to see Rouse, or, more often, disappointment that they have not been able to meet as he had hoped.

At some point before the end of 1737 Dunthorne had been introduced to Roger Long, Master of Pembroke College, Cambridge. In late 1737 Long persuaded Dunthorne to move to Cambridge to work as his servant and assistant. Most biographical accounts of Dunthorne claim that he took this role in order to continue his mathematical education under Long's guidance;[8] it is doubtful that Dunthorne would have learnt much from Long, however. Dunthorne was already at this time a highly competent mathematician, as his contributions to *The Ladies Diary* show, and he had already constructed the first versions of his lunar tables. More likely, Dunthorne was hoping to find fellow astronomers with whom he would be able to discuss his work. In this he would soon be disappointed. Dunthorne arrived in Cambridge on 8 January 1737/8, but within a few weeks he felt he had made a mistake in coming. He wrote to Rouse on 22 March:

P.S. In yours of the 17th. last you desire to know how I like Cambridge, to w[ch]. I answer that I like it tolerably well; but not so, but that I would return to my old business (i.e. teaching schools) again, if I could light on a Place to my mind. So that if you should hear of any Place that way, you would still do me a favour by recommending me to it.[9]

In a letter written to Rouse on 3 June 1738, Dunthorne explained his disillusionment with Cambridge:

I have many things more to say, if I had an opportunity of conversing with you viva voce, and therefore should be very glad to see you at Cambridge, if your affairs would give you leave: I had been at Harborough ere now, but I am a prisoner at Pembroke Hall; and that which renders my confinement tedious is, that the whole University does not afford one practical Astronomer, and scarce three worth the name Theoretical ones; so that I am deprived of that pleasure w[ch]. I had promised to myself, in the enjoyment of such conversation at Cambridge: althou' as you intimate there is no great hopes of encouragem[t]. at Harborough, yet 'tis possible I might be welcomed to some adjacent village and I do assure you I seek only a competency for myself (at present) and more freedom of ingenious conversation.[10]

Dunthorne was greatly disappointed in the level of astronomical knowledge of the University Professors. Robert Smith, second Plumian Professor Astronomy from 1716 to 1760 and director of the observatory at Trinity College, was mainly interested in optics and he made little contribution to astronomical observation or theory. The Lucasian Professor of Mathematics from 1710 to 1739 was the blind Nicolas Saunderson, and although Saunderson gave regular lectures on Newtonian physics, his interests did not extend to astronomy. And as for Roger Long, his employer at Pembroke, Dunthorne felt him to be barely competent and said as much to Rouse in a letter of 25 April 1739:

Professor Saunderson died here last Thursday, and 'tis thought, that (unless the Electors will dispense with my Master, who offers himself as a Candidate, although he does not pretend to understand Algebra or Geometry) they will be obliged to send to Edinburgh for a Person to succeed him. A sad instance of the State of Mathematics here![11]

In the event, the Lucasian chair went to John Colson of Rochester, but like Saunderson, Colson was not interested in astronomy. John Martyn, Professor of Botany from 1733 to 1762, observed the Aurora Borealis in 1749–1750 and reported his observations to the Royal Society, but does not appear to have been interested in astronomical theory. Probably only Charles Mason, the Woodwardian Professor of Geology from 1734 to 1762, who made observations from the Observatory at Trinity College and had several astronomical instruments installed in his rooms at Trinity, might have been

---

[8] For example Clerke and McConnell (2004), Croarken (2007), Skempton (2002).

[9] MS Museum 95, p. 17.

[10] MS Museum 95, p. 64.

[11] MS Museum 95, p. 70.

considered worth discussing astronomy with. Dunthorne's letters to Rouse often end with the statement that "Dr Long and Mr Mason present their service to you", and Dunthorne's letters published in the *Philosophical Transactions* were all addressed to either Long or Mason, so it seems likely that it was with these two that Dunthorne had the most contact.

Dunthorne's complaints about his situation at Cambridge continue in his letters to Rouse for the next few years. He has "no will of my own, but (you will consider me) as being wholly under the power of another",[12] is "a slave to the will of another",[13] and is "compelled to spend a few more unpleasant days in the Captivity".[14] Finally, however, Dunthorne was offered an escape: Long arranged for Dunthorne to be appointed master of the free school in Coggeshall in Essex. Dunthorne joyfully wrote to Rouse on 7–8 January 1741/2 that he would now have more time for astronomy:

> I must desire you would (at your first oppertunity) compute the Moon's Par^x. from the observations of Oct.19.1678 & Feb.11.1682, that I sent you for that purpose, for I am apt to suspect the Moon's Parallax bigger than our Tables make it. I will do the same as soon as I can Procure Time for that purpose, wch I hope I shall be able to do after Lady's Day, I having at last (by many stratagems) procured Dr Long's Consent y^t I shall go at that Time to a School in College's Gift at Great Coggeshall in Essex.
>
> I should not have been so eager to have left Cambridge, but that Dr Long exacts the Meanest Services, & expects the most Slavish Obedience from me; wch I think (considering the worth of my services to him in some other respects) he should not do: nay and I should even have put up with this, if my staying here could possibly be of service either to the Sciences, or my Self; but the Sons of Alma Mater seem to have forsaken Minerva's Temple for those of Bacchus & the Cyprian Queens. If I could have procured the use of the Observatory at Trinity College it would have been a pleasure for me to have stayed here longer; but this I can by no means do, although no other Person makes any use of it: and besides the few leisure hours I have here are not near so Serene as the Contemplative Sciences require.[15]

Dunthorne's time at Coggeshall cannot have gone smoothly, however, for within two years he was back at Pembroke Hall, where he was appointed butler of the college. His letters to Rouse continue to bemoan his treatment by the University, where he had been overlooked for the position of Schoolkeeper. On 3 July 1744 Dunthorne wrote:

> My Time and Thoughts were taken up, for some Time lately in making interest for the Schoolkeeper's Place in this University, then vacant, but to little purpose, for the University preferred an ignorant Sott, set up by men in Power before me: you'll guess how I relished such usage where I might have expected better.[16]

In 1746 Dunthorne married Elizabeth Hill of Huntingdonshire, and after this time he complained less of life in Cambridge. His subsequent letters to Rouse focus on his astronomical work, with occasional asides about purchasing cloth for his wife. By 1750 Dunthorne's astronomical interests were turning away from lunar tables. In the college library at Pembroke, Dunthorne had found a medieval manuscript containing observations of comets. On the basis of similar orbital elements, Dunthorne identified a comet seen in 1264 with one seen in 1556 and concluded that it was periodic with a period of about 292 years. His letter on the subject, addressed to Long, was read to the Royal Society in 1751.[17]

In 1753 Dunthorne was elected Surveyor of the South and Middle Level of the Fens with an initial annual salary of £80. In 1764 he became Superintendent General for the whole of the Bedford Level, in which capacity he supervised the improvement of the outfall of the river Nene below Wisbech.[18] He held the position of Superintendent General of the Bedford Level until his death, combining it with his duties as Butler at Pembroke.

---

[12] MS Museum 95, p. 94.

[13] MS Museum 95, p. 98.

[14] MS Museum 95, p. 112.

[15] MS Museum 95, p. 120.

[16] MS Museum 95, p. 124.

[17] Dunthorne, "Concerning Comets".

[18] Skempton (2002), p. 196.

Dunthorne maintained his interest in astronomy despite the responsibilities of his two careers. He acted as Long's assistant, helping him determine the latitude of Cambridge using a specially designed gnomon attached to the south-east corner of the Pembroke's chapel,[19] and build Long's famous "Great Sphere", an 18-ft-diameter sphere in which more than thirty people could sit and experience the motions of the heavenly bodies.[20] Dunthorne began to work on tables for the motions of the satellites of the planets, and a letter from Dunthorne to Mason on the motion of Jupiter's satellites was read before the Royal Society on 5 March 1761. In 1765 Dunthorne was appointed first comparer of the *Nautical Almanac* by the Astronomer Royal Nevil Maskelyn, and he contributed to the *Nautical Almanac* for 1767. In recognition of his work, Dunthorne, together with Israel Lyons, a computer for the *Almanac*, was awarded £50 by the commissioners of the Board of Longitude for their calculations of the lunar distance.

In 1765 Dunthorne supervised and funded the building of an observatory at St. John's College, equipping it with his own instruments. The observatory was operated first by the Rev. William Ludlam, a former Fellow of St. John's and at that time vicar at Cockfield, and then by the Rev. Thomas Catton, a tutor at the college.[21]

Long's death in 1770 brought a new responsibility to Dunthorne. Long named Dunthorne his chief executor, and the duty fell on Dunthorne to complete Long's *Astronomy in Five Books*, still unfinished more than three decades after the first part had been completed. Dunthorne contributed to the writing of the final book, a history of the development of astronomy from antediluvian times down to the present. However, other duties prevented Dunthorne from devoting the necessary time to this project, and it fell to William Wales to complete the work.

The relationship between Dunthorne and Long has generally been portrayed as one between an eager and grateful Dunthorne and his kindly patron Long. For example, the biographical note appended to the abridged account of Dunthorne's paper on the secular acceleration of the moon published in the ninth volume of *The Philosophical Transactions of the Royal Society of London, From Their Commencement, in 1665, to the Year 1880; Abridged, with Notes and Biographic Illustrations* by Charles Hotton, George Shaw, and Richard Pearson reads:

> … he so distinguished himself, as to gain the notices of his superiors, among whom was Dr. Long, master of Pembroke Hall, Cambridge, who afforded him great encouragement; and at length removed him to Cambridge as his foot-boy. … On the death of his patron, in 1770, by whom he was always treated with the greatest kindness, and with whom he always lived in the strictest intimacy and friendship, he found himself named in the doctor's will, as one of his executors, an office which he discharged with every possible attention.[22]

However, Dunthorne's letters to Rouse paint another picture, where Dunthorne felt exploited by Long, "wholly under the power of another",[23] and "afraid the D^{rs}. ill usage of me, maugre all my services to him"[24] has left him with no time of his own. Not having Rouse's letters to Dunthorne, however, it is hard to judge the seriousness of these remarks. For example, it is possible that over-exaggerating these complaints was a running joke between Dunthorne and Rouse. Nevertheless, Dunthorne was obviously unhappy during his first few years at Pembroke. He may have hoped that Long and the other Cambridge scholars would discuss astronomy with him as their equal, or at least on the same level as a favoured student, but instead found that Long treated him as a servant. Dunthorne's association with Long may have created barriers with some of other members of the University. Long, a tory, came into

---

[19] Gunther (1937), pp. 164–165. See also Dunthorne's letter to Rouse of 5 March 1750/1 (MS Museum 95, pp. 160–163).

[20] Gunther (1937), p. 167.

[21] Mullinger (1901), pp. 243–245; *The Eagle* 7 (1871), pp. 334–337; Gunther (1937), pp. 169–172 and 193–203.

[22] Hotton et al. (1809), pp. 669–670.

[23] MS Museum 95, p. 94 (Dunthorne to Rouse 24 July 1740).

[24] MS Museum 95, p. 119 (Dunthorne to Rouse 12 Aug 1741).

frequent conflict with the whigs at the predominantly whig University,[25] and also had battles with the fellows of his own college over the election of new fellows.[26] We might also speculate that since Dunthorne did not hold a degree in divinity, he may have felt excluded from the company of the ordained fellows and professors. Things improved, however, and by the mid-1740s, especially after his elevation to butler of the college, Dunthorne's complaints about Long cease to be a feature of his letters to Rouse. There may have been several reasons for this. Perhaps in his new role, Dunthorne felt Long finally treated him more as his equal. Quite possibly, Dunthorne's marriage to Elizabeth Hill in 1746 softened Dunthorne's sense of injustice in the world. Dunthorne's standing within the scientific community certainly improved after the publication of his two studies on the lunar theory in the *Philosophical Transactions* in the late 1740s. Astronomers from other parts of Europe discussed Dunthorne's work (including Tobias Mayer, as I will discuss in Chap. 7). When Lalande visited England in 1763, at a meeting with John Michell, Charles Mason's successor as Woodwardian Professor of Geology at Cambridge, Lalande spoke with him about "Smith, Dunthorne, Long",[27] putting Dunthorne in the same category as Robert Smith, Plumian Professor of Astronomy and Natural Philosophy, and Long, then the Lowndean Professor of Astronomy.

Dunthorne died on 3 March 1775 at the age of 64 years, and was survived until 1789 by his wife Elizabeth. In 1773, Dunthorne had been engaged to survey the Stour river in Kent. He conducted the survey at the end of May 1774, but shortly after returning to Cambridge he suffered a "Stroke of the Palsy",[28] and although he recovered sufficiently by the September to send his report to the Commission of Sewers, written as ever in his clear, elegant handwriting (subsequently published by the Commission), he died the following March. Dunthorne's books were sold through the bookseller Samuel Parker of New Bond Street in 1776.[29] According to Hutton, Shaw, and Pearson, Dunthorne left behind "a great number of valuable manuscripts and drawings, most of which were inconsiderately burnt soon after his death, as waste paper".[30]

The Practical Astronomy of the Moon

In 1739 Dunthorne published *The Practical Astronomy of the Moon: or, new Tables of the Moon's motions, Exactly constructed from Sir Isaac Newton's Theory, as published by Dr Gregory in his Astronomy, With Precepts for computing the Place of the Moon, and Eclipses of the luminaries*, printed for the author and to be offered for sale by John Senex in London and James Fletcher in Oxford. The work was dedicated to Roger Long. Despite Dunthorne's unhappiness in his relationship with Long, Dunthorne probably felt he had no option other than to formally acknowledge his obligation to Long. Dunthorne charged 2:6 for copies of the book.

It is clear from Dunthorne's letters to Rouse that he had already more or less completed the construction of his tables before he moved to Cambridge in 1737, but had spent some time checking the tables and developing a new method of computing eclipses of the sun—not to mention arranging the printing of the volume (Dunthorne wrote to Rouse explaining that "the composing of the whole work has gone through my own hands, and that I was an entire stranger to Printing before"[31])—and it was the middle of 1739 before the book was ready. Dunthorne's letters to Rouse indicate that he explored

---

[25]Gascoigne (1989).

[26]Attwater (1936) pp. 94–98.

[27]Monod-Cassidy (1980), p. 37.

[28]MS BL Add. 5489, f. 119.

[29]Nichols (1812), vol. 3, p. 655.

[30]Hotton et al. (1809), p. 670.

[31]MS Museum 95, p. 68. (Dunthorne to Rouse 25 April 1739).

various corrections to the lunar theory, but in his note to the reader he remarks that the tables as published were based purely on Newton's theory:

> The following Tables of the lunar Motions were at first calculated purely for my own private use; but being informed by several friends who are more conversant in Books than myself, that there has not been a complete Set of such Tables framed from the Theory of Gravity, as yet published: I was willing to present you with these: They are constructed from Sir Isaac Newton's Theory, as published by Dr. Gregory in his Astronomy; and though I find, by comparing them with observations, that the Newtonian Numbers are a little deficient, they will (at least) have this use, that such Persons as desire further to rectify the Lunar Astronomy, may be being assisted with Tables already framed from nearly true Numbers, be better enabled to compare those Numbers with Observations, and by that means obtain Numbers still more exact.[32]

Dunthorne was incorrect in stating that his tables were the first constructed from the gravitational theory. At least six sets of tables were compiled and published before Dunthorne's.[33]

Dunthorne's remark that he has compared the tables with observations and found that the "Newtonian Numbers are a little deficient" is confirmed by his correspondence with Rouse. Several of his letters written during 1738 remark on errors with the Newtonian parameters, and explain that he intended to analyse available eclipse observations in order to improve the parameters. He seems to have made only limited progress in this, however, and his tables are a faithful representation of Newton's original lunar theory, except that Dunthorne correctly changed the Equation of the Apogee to the value as calculated from the model.[34]

Following the publication of *The Practical Astronomy of the Moon*, Dunthorne immediately returned to the task of comparing his tables with observations in order to assess and correct the parameters of the lunar theory. He began by investigating the solar theory, for any error in the solar theory would influence the lunar theory. Dunthorne wrote to Rouse on 20 June 1739, only a couple of months after his tables had been published, that he had obtained observations by al-Battānī from which he deduced the rate of precession and the obliquity of the ecliptic:

> I have lately seen the small remains of that celebrated Astronomer of Antiquity, Albategnius the Arabian, called also Mahomet Aractensis, entitled De Scientia Stellarum, who says, that with a Parallactic Instrument of Ptolemy, whose side and Alhidase or sight-ruler were very long, he carefully observed (for several years) the least distance of the Sun from the Zenith of Aracta to be 12°26', and its greatest Meridional distance 59°36', whence he collects the distance of the Tropics 47°10', obliquity of the circle of the Signs i.e. Ecliptic 23°35' and the latitude of Aracta 36° he likewise says that Anno ad Hilkarnain 1191 i.e. ab Alexandri morte 1203 he carefully observed, by the passage of ☽ and Stars over the meridian, that the Northern Stars in the Forehead of the Scorpion was in ♏ 17°20' and the Lyons heart in ♌ 14'.
> By comparing these places wth. the places of the same Stars in Flamsteed's Catalogue I find the Precession is 11°32' in 810 years w[ch]. is 51″ & a little more in a year: and from his observations of the Sun's distance from the Zenith, I think it is almost certain the inclination of the Ecliptic and the Equator must have changed since his Time, though I think it as certain it has not sensibly altered since Bernard Walthers Time: I should be glad if some Astronomers would undertake a journey to Aracta, to see how its Latitude agrees with those observations, and also to take its Longitude from Greenwich, that his Observations of Eclipses might be of more use than they can be without it. I have more things to communicate when I see you.[35]

By the following January, Dunthorne had turned to the lunar theory proper and came to the conclusion that the epochs of the mean longitude of the moon are about 1' more than in his tables. To investigate further, Dunthorne needed more observations. On 2 October 1740, he wrote to Rouse that

> I now have in my possession, and can procure, Ptolemy's Almagest, Albategnius de Scientia Stellarum, The observations of Regiomonanus & Bernard Walther, Copernicus de Revol., The observations of the Prince of

[32] Dunthorne, *The Practical Astronomy of the Moon*.

[33] Waff (1977), Kollerstrom (2000), p. 208.

[34] Kollerstrom (2000), p. 94. Kollerstrom comments that "Dunthorne's 1739 opus appears to me as the one work which has embodied 100% the *TMM* (Newton's *Theory of the Moon's Motion*) rules in its lists of tables and instructions on how to use them" (p. 99).

[35] MS Museum 95, pp. 77–78.

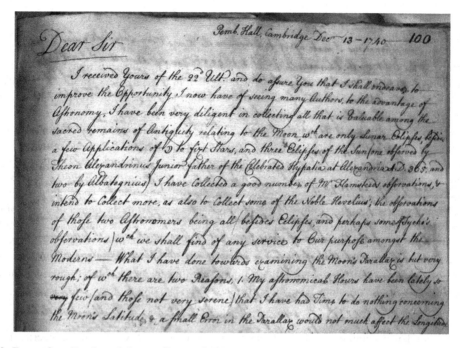

Fig. 6.2 Extract from Dunthorne's letter to Rouse of 13 December 1740 (MS Museum 95, p. 100) (Courtesy Museum of the History of Science, Oxford)

Hesse, Gassendus & Ricciolus, Those of Horrox & Crabtree, and all the observations have been published of those great modern Astronomers Tycho Brahe, Hevelius & Flamsteed (Hevelius' observations of the Moon are over 2000, and Tycho's are a good number) I have also several Observations made at yᵉ French Royal Observatory, so that I persuade myself I shall be able to furnish almost what will be wanting on that head.

Already by this time Dunthorne recognised the value of ancient observations. Small inaccuracies in the parameters of the lunar theory might accumulate over long timescales and so be easily apparent from comparison of theory with even rough observations made in antiquity. Dunthorne remarked in his next letter to Rouse, dated 13 December 1740, that "I have been very diligent in collecting all that is valuable among the sacred remains of Antiquity relating to the Moon, wᶜʰ are only Lunar Eclipses, besides a few Applications of ☽ to fixt Stars, and three Eclipses of the Sun (one observed by Theon Alexandrinus Junior father of the Celebrated Hypatia at Alexandria A.D. 365 and two by Albategnius)"[36] (Fig. 6.2).

Unfortunately for Dunthorne, his duties at Pembroke Hall were by now taking up almost all of his time. Although in the summer of 1741 Dunthorne finally managed to visit Rouse in Market Harborough, his work as Long's footman, and his subsequent teaching duties at the school in Great Coggeshall meant that over the next few years he could devote very little time to astronomy. Indeed, his correspondence with Rouse, at least as preserved in Rouse's collection, was reduced to only one letter between 1741 and 1748. Nevertheless, by 1746 Dunthorne had successfully completed his analysis of the more recent lunar and solar observations (largely confirming his preliminary findings from 1739 to 1740) and wrote a letter on the subject to Charles Mason dated 4 November 1746 which was read before the Royal Society on 5 February 1746/7.[37]

---

[36] MS Museum 95, p. 100.

[37] Dunthorne, "Concerning the Moon's Motion".

Dunthorne began his letter with a discussion of the solar theory since "As the motion of every secondary Planet must partake of the Errors in the Theory of its primary, I thought proper, before I undertook the Examination of the lunar Numbers, to compare those of the Sun with Observations". He reported that he has compared several sets of observations of the sun's position made by Flamsteed with his tables, with the help of "a Gentleman well skilled in these Matters" (almost certainly a reference to Rouse, who Dunthorne had asked to check his findings in a letter dated 19 January 1739/40). Dunthorne found that the mean motion of the sun on 31 December 1700 (old style) at noon in Greenwich is 20°43′40″ in Capricorn, its apogee 7°30′0″ in Gemini, and the greatest equation of the sun's centre 1°55′40″, as compared with 20°43′50″ in Capricorn, 7°44′30″ in Gemini and 1°56′20″ respectively given in the first version of Newton's lunar theory which underlies Dunthorne's tables. He remarks that these results are "very near the Truth", although in his 1739/40 letter to Rouse, Dunthorne had noted that there seems to be a greater disagreement between the sun's position deduced from the transit observations that "could possibly arise from the defect of observations of so accurate a man as M^r. Flamsteed",[38] and wondered if the gravity of the planets somehow affected things. Unfortunately, as we do not have Rouse's reply, we do not know whether Rouse was able to reassure Dunthorne that the discrepancies in the solar positions deduced from Flamsteed's transit observations were insignificant, or whether Dunthorne himself either came to this conclusion or simply forgot about the problem. The value for the greatest equation of the Earth's centre given in Dunthorne's Royal Society letter of 1746/7 is identical to that in his letter to Rouse of 1739/40 suggesting that Dunthorne's discussion of the solar theory was based purely upon his earlier study.

The remainder of Dunthorne's 1746/7 letter published by the Royal Society concerns the lunar theory. As he had already told Rouse in January 1739/40, from comparison with observations, Dunthorne found that the moon's mean longitude should be about 1′ greater than given in his tables, in close agreement with the value published by Newton in his revision of the lunar theory in the final edition of the *Principia*. In order to examine the longitude and motion of the apogee, the theory of the increase and decrease of the eccentricity and the values of the greatest and least eccentricities, Dunthorne reported comparisons of his tables with 100 observed longitudes of the moon: 25 eclipses of the moon dating from 1652 to 1732, all but 1 taken from Flamsteed's *Historia Coelestis*, the *Philosophical Transactions* and the *Memoirs of the Royal Academy of Sciences*; 2 observations of solar eclipses seen in 1706 and 1715, 25 lunar longitudes calculated by Dunthorne from observations given in Flamsteed's *Historia Coelestis* dating from 1684 and 1693 to 1714; and 48 lunar longitudes dating from 1689 to 1691 computed from Flamsteed's observations by Halley and printed in the bootleg edition of the *Historia Coelestis*. For four of the observations, the earliest lunar eclipse observation from Hevelius and three of the lunar longitudes calculated by Halley, Dunthorne noted errors (either computational or printing) in his sources and made the appropriate corrections. It is interesting that Dunthorne confines himself to these 100 observations in his Royal Society letter. In a letter to Rouse dated 12 August 1741, Dunthorne tests his tables against another eclipse observation made on 10 January 1647 by Hevelius,[39] and in his earlier letter dated 2 October 1740 quoted above, Dunthorne had remarked that he had access to the observations of Horrocks, Crabtree, and Hevelius (whose lunar observations alone Dunthorne says number over 2,000),[40] all of whom were experienced and competent observers. It is not known whether Dunthorne simply did not have the time to analyse all of these observations (lack of time being his constant complaint during this period), or whether he felt that they were of insufficient quality to be of use in refining his tables, or

---

[38] MS Oxford p. 88.
[39] MS Museum 95, pp. 116–119.
[40] MS Museum 95, pp. 96–98.

indeed whether he did analyse them but did not feel it necessary to include the analysis (which even presented in tabular form would have taken several pages) in his letter for publication.

On the basis of his comparisons between observation and his tables, Dunthorne proposed only small corrections to the 6th equation of Newton's theory, and a 1' increase in the mean longitude of the moon. In his subsequent work, Dunthorne always used his corrected tables incorporating these changes.

Dunthorne's Investigation of the Secular Acceleration of the Moon

In his 1746/7 letter published in the *Philosophical Transactions* Dunthorne restricted his comparisons between observed lunar positions and data generated from his lunar tables to fairly recent observations (principally those of Flamsteed). However, as early as 1739, Dunthorne had recognised the value of ancient observations in testing the lunar theory, and it was to these ancient eclipses that he now turned his attention. On 10 March 1748/9 Dunthorne wrote to Rouse that "What leisure Time I have been able to procure from the necessary Occurrences of Life, has lately been sent in comparing ancient Eclipses with the Tables" but he has now "quite finished what I intended about these Eclipses".[41] Indeed, only a week and a half earlier, on 28 February, Dunthorne had written a letter to Charles Mason to be read at the Royal Society reporting his study of the ancient eclipses and announcing that he had been able to confirm Halley's claim that the moon was subject to a long-term acceleration in its motion and providing the first determination of the magnitude of that acceleration.

It is interesting that Dunthorne did not discuss his work on the secular acceleration with Rouse. In their earlier correspondence, Dunthorne had used Rouse as a sounding board for his ideas and frequently asked Rouse to check his calculations. Why did he not do so now? Dunthorne even began his letter to Rouse with the statement "I received your favour of the 4th Instant; and am sorry to inform you that I know of nothing new going on here either Mathematical or Philosophical".[42] Did Dunthorne believe that his work on the secular acceleration was so important that he did not want to share it with Rouse? Other astronomers around this time were beginning to pay attention to Halley's claim that there was a secular acceleration. In particular, George Smith in his *Dissertation on Eclipses* of 1748 presented the secular acceleration as if it were established fact. Perhaps Dunthorne felt that he was going to be pipped at the post to be the first person to firmly establish the existence of the acceleration of the moon and to deduce its magnitude, after many frustrating years in Long's service unable to find the time to complete his work.

Dunthorne's letter to Mason concerning the moon's acceleration was read at the Royal Society on 1 June 1749, and printed in volume 46 of the *Philosophical Transactions* under the title "A Letter from the Rev. Mr. Richard Dunthorne to the Reverend Mr. Richard Mason F. R. S. and Keeper of the Woodwardian Museum at Cambridge, concerning the Acceleration of the Moon".[43] The title contains two typographical errors: Dunthorne is referred to as the "Rev. Mr. Richard Dunthorne", but Dunthorne was never ordained a priest;[44] and Charles Mason is called "Richard Mason". Such errors were not uncommon in the printing of the *Philosophical Transactions* and, as we will see, an error also appears in the date of one of the eclipses discussed by Dunthorne. The letter itself is brief, taking up only 11 printed pages in the *Philosophical Transactions*, and does not contain full details of his analyses. However, enough details are given to allow anyone familiar with the topic (and with Dunthorne's tables and his earlier letter) to follow his arguments and check his calculations.

---

[41] MS Museum 95, p. 142.

[42] MS Museum 95, p. 142.

[43] Dunthorne, "Concerning the Acceleration of the Moon".

[44] This error has been repeated in almost all references to Dunthorne by modern scholars.

The basis of Dunthorne's investigation of the secular acceleration of the moon was to compare observations of eclipses throughout history with the computed circumstances of those eclipses given by his tables. In his letters to Rouse of 2 October and 13 December 1740, Dunthorne mentioned observations from Ptolemy's *Almagest*, Theon of Alexandria, al-Battānī's *De Scientia Stellarum*, Regiomontanus, Bernard Walther, Copernicus's *De Revolutionibus*, Tycho Brahe, the Landgrave of Hesse, Gassendus, Riccioli, Horrocks, Crabtree, Hevelius, Flamsteed, and observations from the French Royal Observatory. It is interesting to compare this list with the eclipses used by Dunthorne in his final published study of 1749. In the order in which they are presented in his work (roughly reverse chronological), these observations are those of Tycho Brahe, Regiomontanus and Walther, al-Battānī, Ibn Yūnus, Theon, and three from Ptolemy's *Almagest*. Dunthorne had used the observations by Hevelius and Flamsteed in his 1746/7 study and those of Gassendus, Riccioli, Horrocks, and Crabtree were probably considered by Dunthorne to be too recent to examine the long-term behaviour of the moon's motion. However, the observations by William IV, Landgrave of Hesse are contemporary with those of Tycho, and Copernicus's are earlier. Why did he not use those? In the case of Copernicus's eclipse observations Dunthorne probably felt that they were too imprecise to be of use. Copernicus gave the time of the eclipses he observed in hours and fractions of an hour. The greatest precision he gave was to the nearest twelfth of an hour, but often the precision is no more than a fifth or an eighth of an hour.[45] This implies a precision of at best about 8 minutes of time. By contrast, Regiomontanus and Walther, who observed only a few years prior to Copernicus, often recorded the times of the various phases of the eclipse to the nearest minute, frequently supplementing the time with the altitude of either the moon or a star at that moment (from which the time had usually been deduced).

Dunthorne remarked in 1740 that he had been "very diligent" in searching for eclipse observations. The starting point for his search may well have been the eclipse catalogue contained in Riccioli's *Almagestum Novum* or his *Astronomia reformata*. This catalogue contained summaries of all the observations up to the time of Tycho mentioned by Dunthorne in his 1740 letters to Rouse. In addition, Riccioli listed many observations of solar and lunar eclipses reported in classical and medieval works. It is not known whether or not Dunthorne trusted such accounts; they were irrelevant for his study as they generally did not contain sufficient detail to determine the time of the eclipse which was essential in fixing the longitude of the moon at that moment.

The eclipses observed by Ibn Yūnus first became known in Europe through Golius's manuscript of Ibn Yūnus's *zīj*, Latin translations of excerpts of which were published by Curtius in the *Historia Coelestis*. The absence of Ibn Yūnus's name from the list of astronomical observations given by Dunthorne in his 1740 letters to Rouse indicates that Dunthorne was not able to consult Curtius's book until after this date.

Dunthorne may have used Riccioli's and perhaps Curtius's catalogues of historical eclipse observations as a starting point for his investigations, but it appears that he tried to use the original sources wherever possible to provide the accounts of the eclipses that he would use in his analyses. For Tycho's observations, Dunthorne said in his Royal Society letter that he has taken them from Tycho's *Progymnasmata*. The *Astronomiae Instauratae Progymnasmatum* was published in three parts in Prague in 1602 (the year after Tycho's death). Page 114 of the first part of this work has a table summarising 21 lunar eclipses and 9 solar eclipses observed by Tycho between 1573 and 1600.[46] Tycho gave for each eclipse the moment of mid-eclipse and an estimate of the magnitude of the eclipse. Dunthorne remarked that Tycho must have experienced some difficulty in determining the times of mid-eclipse from his raw observations as Curtius had presented them in the *Historia Coelestis*. When

[45]Comparison of Copernicus's eclipse timings with modern theory shows that Copernicus's observations were considerably less accurate than those of Regiomontanus and Walther. See Steele (2000), p. 150.

[46]Dreyer (1913–1929), vol. 2, p. 98.

observing eclipses, Tycho often recorded the extent of the eclipse shadow at timed moments during the eclipse. Sometimes they were accompanied by a sketch of the eclipse shadow. The observations do not always give an explicit measurement of the time of the various eclipse contacts and so often the moment of mid-eclipse could only be estimated from the available data. Nevertheless, Tycho's times of the middle of the lunar eclipse given in the *Progymnasmatum* agree well with modern computation, more than three quarters of them agreeing to within 6 minutes of time although with a tendency to be early. Dunthorne did not give his calculations for Tycho's eclipses, remarking only that they "agree full as well as could be expected considering the Imperfection of his Clocks",[47] especially given the difficulty in defining mid-eclipse based upon Tycho's observational records; in any case "the small Distance of Time between *Tycho Brahe* and *Flamstead* render'd *Tycho*'s Observations but of little Use in this Enquiry".

Dunthorne next considered the eclipses observed by Bernard Walther and Regiomontanus. Between 1457 and 1471 Regiomontanus observed eight lunar and one solar eclipse in Melk, Vienna, Rome, Viterbo, Padua, and Nuremberg (the first two eclipses were observed with the help of Regiomontanus's teacher Georg Peurbach). Bernard Walther observed four solar and two lunar eclipses in Nuremberg between 1478 and 1504. Regiomontanus and Walther's eclipse observations were published by Schöner in the *Scripta Clarissimi Mathematici M. Ioannis Regiomontani*.[48] The eclipses reported in Schöner's edition were included in Curtius's *Historia coelestis* and Riccioli's *Almagestum novum* and *Astronomia reformata*. It is not known whether Dunthorne consulted Schöner's edition directly or the quotations of the eclipse accounts given by Curtius and Riccioli. As with the eclipses observed by Tycho, Dunthorne did not give any details of his analysis of Regiomontanus and Walther's eclipses, remarking only:

> Upon comparing such of their Eclipses of the Moon whose circumstances are best related with the Tables, I found the computed Places of the Moon were mostly 5' too forward, and in some considerably more, which I could hardly persuade myself to throw upon the Errors of Observation; but concluded, that the Moon's mean Motion since that time, must have been something swifter than the Tables represent it; thought the Disagreement of the Observations between themselves is too great to infer any thing from them with Certainty in so nice an Affair.[49]

Regiomontanus and Walther generally obtained the times of the eclipses they observed by measuring the altitude of either a star or the eclipsed body. Comparison of their observations with modern computation has shown that the mean accuracy of their altitude measurements is about 0.8°, corresponding to an error in time of just over 7 minutes, but with some errors up to about 20 minutes.[50] These correspond to errors in lunar longitude on an average of about 3', but sometimes as much as about 9'. Dunthorne's claim of a 5' systematic discrepancy in the longitude of the moon between his tables and the observations is of the same order of magnitude as the random error in the observations, so Dunthorne was correct in concluding that these observations did not provide conclusive proof of

---

[47]Dunthorne, "Concerning the Acceleration of the Moon", p. 162. Tycho used a mixture of mechanical clocks and observations of stellar altitudes to time the eclipses. For example, his report of the lunar eclipse on 26 September 1577 gives times measured by a clock and times deduced from the altitude of α Orion, noting that in this case "the observation by the shoulder of Orion (α Orion) is not very good. In fact, errors of more than 2° may be on account of a poorly defined meridian line, and also may be caused by the wood of the instrument. I prefer to trust the clock." (Dreyer (1913–1929), vol. 10, p. 50).

[48]Regiomontanus and Walther's eclipse observations are given in folios 36–43, entitled "Ioannis de Monterehio, Georgii Peurbachii, Bernardi Waltheri, ac aliorum, Eclipsium, Comentarum, lanetarum ac Fixarum observationes", and folios 44–60, entitled "Observationes factae per doctissimum virum Bernardum Waltherum Norimbergae", respectively. English translations and a study of the accuracy of the times of these eclipses are given in Steele and Stephenson (1998).

[49]Dunthorne, "Concerning the Acceleration of the Moon", p. 163.

[50]Steele and Stephenson (1998), Steele (2000), pp. 139–147.

the existence of the secular acceleration, although they were compatible with such an assumption if it could be proven from other data.

Dunthorne next considered the four eclipses reported in al-Battānī's *De Scientia Stellarum*. Once more, these eclipses seemed to suggest the existence of a secular acceleration, but did not provide conclusive proof. This time an additional uncertainty came from lack of knowledge of the exact location of the city of Aracta (al-Raqqa) from which they were observed:

> Then I compared the four well known Eclipses observed by *Albategnius* with the Tables, and found the computed Places of the Moon in three of them considerably too forward: This, if I could have depended upon the Longitude of *Aracta*, would very much have confirmed me in the Opinion, that the Moon's mean Motion must have been swifter in some of the last Centuries than the Tables make it; though the Differences between these Observations, and the Tables, are not uniform to be taken for a certain Proof thereof.[51]

As was remarked by Halley (see Chap. 2), considerable variation may be found in the coordinates of cities in the near east given in different seventeenth- and eighteenth-century tables of geographical latitudes and longitudes. Depending upon which set of coordinates one used, it was possible either to show either that there was no need for a secular acceleration or that there must be one. Like Halley before him, Dunthorne had 10 years previously expressed the wish that "some Astronomers would undertake a journey to Aracta, to see how its Latitude agrees with those observations, and also to take its Longitude from Greenwich, that his Observations of Eclipses might be of more use than they can be without it",[52] but no one had undertaken the task.

Thus far, Dunthorne had been frustrated in his efforts to investigate whether the moon is subject to a secular acceleration: the eclipses observed by Tycho, Regiomontanus and Walther, and al-Battānī all suggested that there is an acceleration, but in each case the uncertainties in the timing of the eclipses or knowledge of the geographical coordinates of the place of observation meant that he could not be sure of this conclusion. Next, Dunthorne turned to the very few other eclipse records from between the time of Ptolemy and Regiomontanus, which are only, he says, two solar eclipses and one lunar eclipse observed in Cairo, found in the manuscript of Ibn Yūnus and translated in Curtius"s *Historia Coelestis*, and the solar eclipse observed by Theon of Alexandria. Dunthorne remarked that:

> These eclipses of the Sun are the more valuable, because they were observed in Places the Longitudes and Latitudes whereof are determined by Monsieur *Chazelles* of the *Royal Academy of Sciences*, who was sent by the *French* King in the Year 1693, with proper Instruments for the Purpose. *Du Hamel Hist. Acad. p.* 309, 395.[53]

Jean Mathieu de Chazelles (1657–1710) was a professor of hydrology at Marseilles and an experienced surveyor.[54] In 1693, Chazelles was sent on an expedition to survey Egypt and parts of the Mediterranean. In Egypt, Chazelles measured the latitude and longitude of Alexandria and Cairo, and surveyed the pyramids at Giza, noting that the sides of the pyramids were very precisely aligned with the cardinal directions. Chazelles's survey of Egypt was communicated to the Académie Royale des Sciences through its secretary Jean-Baptiste Du Hamel, and was used by Cassini and other members.

Dunthorne's reference to "*Du Hamel Hist. Acad. p.* 309, 395" for Chazelles's coordinates of Cairo and Alexandria is unfortunately ambiguous. Du Hamel, as secretary of the Académie at the time of Chazelles's expedition, was responsible for the Académie's publications, but the annual *Histoire de l'Académie Royale des Sciences* only began in 1699. It is likely that Dunthorne was referring to the later collection of material *Histoire de l'Academie Royale des Sciences Depius 1686 jusqu'a son*

---

[51] Dunthorne, "Concerning the Acceleration of the Moon", p. 163.

[52] MS Museum 95, p. 78.

[53] Dunthorne, "Concerning the Acceleration of the Moon", p. 164.

[54] Sturdy (1995), pp. 257–259.

Renouvellement en 1699, published in 1733 many years after Du Hamel's death. This volume was published in at least two editions that year, with different pagination. Neither edition that I have been able to consult has any reference to Chazelles on pages 309 or 395. However, the two versions do refer to his determination of the longitude of Alexandria and Cairo on pages 142 and 203 of one edition, and pages 228 and 325 of the other edition. I suspect that Dunthorne read a third edition of this work which had the relevant passages on pages 309 and 395. There is a further complication, however. In the earlier passage, Chazelles is reported to have obtained a difference in longitude between Paris and Alexandria of 1h51′13″ and between Paris and Cairo of 1h58′20″. However, in the second passage, the difference between Paris and Alexandria is given as 2h52′. This second value is clearly a typographical error for 1h 52′ (the modern value is 1h50′30″).[55]

Dunthorne first discussed the record of the solar eclipse reported by Theon. Dunthorne described the eclipse as follows:

> The solar Eclipse observed by *Theon* was in the 112th Year of *Nabonassar* the Day of *Thoth*, according to the *Egyptians*, but the 22nd Day of *Pauni*, according to the *Alexandrians*: He carefully observed the Beginning of 2 temporal Hours and 50′ Afternoon, and the End at 4½ Hours nearly Afternoon at *Alexandria*. *Thenonis Comment. In Ptol. Mag. Construct.* p. 332. This Eclipse was *June* 16, in the Year of Christ 364: And the temporal Hour at *Alexandria* being at that time to the equinoctial Hour as 7 to 6, makes the Beginning at 3 equinoctial Hours and 18′ Afternoon, and the End at 5 equinoctial Hours 15′ nearly.[56]

This description begins with a typographical error in the date of the eclipse: "112th Year of *Nabonassar*", should read "1112th Year of *Nabonassar*". Dunthorne's source for the record of this eclipse was the 1538 Basel edition of the Greek text of Theon edited by Gryneus and Camerarius and appended to their edition of Ptolemy's *Almagest*, where the eclipse is indeed reported on page 332. The eclipse had been listed by both Riccioli and Curtius in their catalogues, but in both cases the date was incorrectly given as A.D. 365. Dunthorne gave the correct date: 16 June 364 A.D. Interestingly, Dunthorne, without comment, interpreted the time of Theon's observations to be in temporal or seasonal hours, which need to be converted into equinoctial hours. However, the Basel edition clearly states that the times are given in equinoctial hours. At first sight, therefore, it would appear that Dunthorne has misinterpreted the Theon's timings.[57] But, in fact, it is the Basel edition that is in error, as Rome has recently established.[58] The manuscript tradition of Theon's commentary is confused at this point: some manuscripts have "equinoctial hours", others simply "hours". However, Theon's reason for including this eclipse observation in his commentary is to show the accuracy of Ptolemy's eclipse theory, and only if we assume that the time is given in seasonal hours is there the agreement between the observation and Theon's calculations that he claims. Dunthorne was correct to interpret Theon's times as measured in seasonal hours. It is very unlikely that Dunthorne had access to any of the medieval manuscripts of Theon, and so it seems probable that Dunthorne came to this conclusion through the same reasoning as Rome, by looking at Theon's comparison of this observation with his calculations.

Dunthorne next turned to the three eclipses known from Golius's manuscript of Ibn Yūnus. In each case, Dunthorne quoted Schickard's Latin translation as published by Curtius. The first report concerned the solar eclipse of 13 December 977. According to Schickard's translation, the eclipse began

---

[55]The eighteenth volume of the Panckoucke edition (I have been unable to check the first edition) of the *Description de l'Égypte*, p. 389 gives Chazelles's value for the longitude of Alexandria as 47°56′33″. This value agrees with neither of the two values given in the *Histoire de l'Académie Royale des Sciences*. However, if we assume the value in the *Description de l'Égypte* is a typographical error for 27°56′33″, this equals 1h51′46″, which is very close to the first value found in the *Histoire de l'Académie Royale des Sciences*.

[56]Dunthorne, "Concerning the Acceleration of the Moon", p. 164.

[57]Riccioli, *Almagestum Novum*, p. 369 followed by Curtius, *Historia Coelestis*, p. xxv, gives the times simply as "hora post meridiem", which could be read as either seasonal or equinoctial hours. Only occasionally does either author distinguish between the two types of hours, even when one type of hour is specified in the original source.

[58]Rome (1950). See also Jones (2012).

when the sun had an altitude of 15°43′ and ended when the solar altitude reached 33½°.[59] Dunthorne converted these into local apparent times of 8h25′ and 10h45′ in the morning. The second eclipse was the solar eclipse of 8 June 978. From Schickard, Dunthorne read that the eclipse began when the solar altitude was 56° and ended when it was 26°,[60] which he converted to local apparent times of 2h31′ and 4h50′ in the afternoon. The final eclipse was the lunar eclipse of 14 May 979, but Dunthorne noted that "as the Middle cannot be known from what was observed of it, I made no use thereof in this Enquiry".[61]

Comparing the longitude of the moon calculated for the observed time of a lunar eclipse with the moon's longitude calculated for the computed time of the opposition, as Dunthorne had done for the eclipses observed by Regiomontanus, Walther, and Tycho, provided an easy method to determine whether the moon had been accelerated in its motion. However, solar eclipse observations could not be used in the same way because (1) the track of a solar eclipse only covers part of the Earth's surface, and (2) the effect of parallax. Therefore, Dunthorne could not analyse the three solar eclipses from Theon and Ibn Yūnus using the simple method he had applied to the lunar eclipses. Dunthorne's solution was to treat the problem geometrically by projecting the position of the sun and moon onto the ecliptic at the observed beginning and end of the eclipse and deducing the distance of the moon from the sun along the ecliptic at the time of mid-eclipse and comparing it with the computed distance of the moon from the sun from his tables.

From the three solar eclipses, Dunthorne found the difference between the moon's longitude deduced from the observations and from his tables to be −4′16″ for Theon's eclipse if A.D. 364, +7′36″ for Ibn Yūnus's eclipse of A.D. 977 and +8′45″ for the eclipse of A.D. 978. Dunthorne takes the agreement between the differences in longitude deduced for the two eclipses observed by Ibn Yūnus to be evidence that his tables determine the mean motion of the moon's apogee well, since the moon was near perigee at one eclipse and near apogee near the other.[62] This implied that the differences between the calculated and observed longitudes are due to a secular acceleration of the mean motion of the moon.

Finally, Dunthorne considered the eclipses in Ptolemy's *Almagest*.[63] But, Dunthorne said, they "are most of them so loosely described, that, if they shew us the Moon's mean Motion has been accelerated in the long Interval of Time since they happened, they are wholly incapable of shewing us, how much that Acceleration has been".[64] The inaccuracies and ambiguities in the timings given in Ptolemy's eclipse reports had been highlighted by Bullialdus and used by Struyck to argue that there was no evidence for a secular acceleration. Dunthorne, however, came up with an innovative way to avoid the problem. Instead of using the times of the eclipses, Dunthorne considered their visibility relative to either sunrise or sunset. For example, if the moon set whilst eclipsed, but calculation using tables put the beginning of the eclipse after moonset, this implied that the longitude of the moon was greater than

---

[59] A recent edition and translation of Ibn Yūnus's eclipse records by Said and Stephenson (1997), p. 37 gives the solar altitude at the beginning and end of this eclipse as "more than 15° but less than 16°" and "more than 33° by about a third of a degree". It is not clear where Schickard obtained the figures in his translation. The error introduced into Dunthorne's calculations was trivial.

[60] The solar altitudes are in agreement with those given by Said and Stephenson (1997).

[61] Dunthorne, "Concerning the Acceleration of the Moon", p. 165.

[62] Mercier (1994), p. 198 misunderstands Dunthorne's remark as implying that the mean motion of the moon agrees well in Ibn Yūnus's time with Dunthorne's tables, but Dunthorne is here talking about the motion of the apogee, not the mean motion. Mercier therefore claims that Dunthorne believes the cumulative effect of the secular acceleration is zero in Ibn Yūnus's time, but this is not Dunthorne's conclusion. Dunthorne claims that a positive correction to the moon's calculated longitude is needed at the time of Ibn Yūnus and a negative one at the time of Theon, and later estimates that the zero correction will be at A.D. 700, 300 years earlier than Ibn Yūnus.

[63] Dunthorne probably used the 1538 Basel edition of Ptolemy as this was printed with the edition of Theon's commentary which he referred to earlier in his paper.

[64] Dunthorne, "Concerning the Acceleration of the Moon", p. 169.

that given by tables. The simple observation of the eclipse's visibility put constraints upon the size of the correction needed, and indicated that the moon must have been accelerated in its motion. This is a valid technique for putting limits upon the size of the secular acceleration on a given date and demonstrates Dunthorne's ingenuity in this new field. This method of using horizon eclipse observations is still used in studying the variable rate of rotation of the Earth.[65]

Dunthorne selected three of Ptolemy's eclipse records as suitable for analysis using his horizon method. First, Dunthorne discussed the eclipse of 22 December 383 B.C. Dunthorne's account of this eclipse mistakenly gives the year as 313 B.C., but this is simply a typographical error,[66] not a mistake by Dunthorne.[67] Full moon in December 313 B.C. took place on the 28th, not the 22nd, and so Dunthorne could not have calculated the circumstances of an eclipse on 22 December 313 B.C. Furthermore, Dunthorne's calculations of the eclipse are what would be expected for the correct date. The eclipse of 22 December 383 B.C. was useful because the eclipse was said to have begun half an hour before sunrise and the moon set eclipsed. According to Hipparchus, the eclipse was observed in Babylon, which Dunthorne takes (following Ptolemy) to be 50′ east of Alexandria. From his tables, Dunthorne calculated that middle of the eclipse was at an apparent local time of 9h4′ (printed in error as 4h4′), and the eclipse lasted for 1h37′, so the beginning of the eclipse should have been at about 8h15′ after midnight. But, according to Ptolemy, the length of night at Babylon at this time (almost exactly the winter solstice) was 14h24′, and therefore sunrise would have been at 7h12′ am. Ptolemy's estimate of the latitude of Babylon, based upon the canonical value of 3:2 for the ratio of the longest to the shortest night at Babylon, is slightly too high, but this would only have affected the time of sunrise by a few minutes. Dunthorne notes that the moon would have set a little before sunrise because the moon was at a southerly latitude and had not quite reached opposition. Thus, the moon must have set

more than a hour before the Beginning of the Eclipse, according to the Tables; whereas the Moon was seen eclipsed some Time before her Setting; which, I think, demonstrates that the Moon's Place must have been forwarder, and consequently her Motion since that Time less than the Tables make it by about 40′ or 50′.[68]

This provided the clearest evidence so far that moon's motion must be an accelerated.

Dunthorne used Ptolemy's record of the eclipse of 22 September 201 B.C. to confirm the existence of the acceleration. This eclipse was observed in Alexandria and according to Ptolemy began half an hour before the moon rose. From his tables Dunthorne calculated that the eclipse should have begun at 7h44′ and had a duration of 3h4′, and so it would have begun at 6h12′, about 10′ after the moon rose. If the record was correct in saying that the eclipse began about half an hour before the moon rose (presumably based upon estimating the fraction of the moon's surface covered at moonrise), this would mean that the calculated time was about 40′ later than observed, which translates into about 20′ of lunar longitude.

Finally, Dunthorne used the oldest eclipse reported by Ptolemy: the lunar eclipse of 19 March 721 B.C. observed in Babylon. Ptolemy's account of time of this eclipse is rather imprecise. Toomer translates Ptolemy's words as "The eclipse began, it says, well over an hour after moonrise"[69]; Dunthorne rendered the time as "when one Hour after her Rising was fully past", and contributed "if, by reason of the Latitude of the Expression, it may be not a direct Proof of the Acceleration, it may nevertheless help to limit its Quantity".[70] Rather than trying to deduce an exact time of the observed beginning of the eclipse from these words, Dunthorne instead assumed that the eclipse began at least an hour after

---

[65] Stephenson (1997), pp. 76–79 and 86–89.

[66] Britton (1992), p. 62.

[67] As incorrectly claimed by Mercier (1994), p. 198.

[68] Dunthorne, "Concerning the Acceleration of the Moon", pp. 169–170.

[69] Toomer (1984), p. 191.

[70] Dunthorne, "Concerning the Acceleration of the Moon", pp. 170–171.

Table 6.1 Summary of the results of Dunthorne's analysis of historical eclipse observations

Source	Date	Difference between observed and calculated longitude
Tycho Brahe	c. 1580 A.D.	Agree full as well as could be expected
Regiomonatus and Walther	c. 1480 A.D.	Computed Places of the Moon were mostly 5′ too forward
Ibn Yūnus	978 A.D.	Observed—calculated longitude = +8′45″
Ibn Yūnus	977 A.D.	Observed—calculated longitude = +7′36″
al-Battānī	c. 900 A.D.	In three of them (out of four) considerably too forward
Theon	364 A.D.	Observed—calculated longitude = −4′16″
Ptolemy (Babylon)	721 B.C.	Precede … but little more than 50′ at that Time
Ptolemy (Babylon)	383 B.C.	Less than the Tables make it by about 40′ or 50′
Ptolemy (Alexandria)	201 B.C.	Near 20′ in the Moon's Place

moonrise to obtain an upper limit for the value of the difference between the observed and the calculated time. From his tables, Dunthorne calculated that the apparent time of the beginning of the eclipse at Babylon was 8h32′ afternoon. The apparent time of the rising of the moon was at about 5h46′ afternoon and so the observed beginning of the eclipse was at earliest 6h46′. Therefore, the difference between the calculated and observed time of the eclipse was no more than 1¾ hours and so the difference in longitude of the moon could be "but little more than 50′ at that Time".[71]

From his analysis of historical eclipse records, Dunthorne could conclude that the moon's motion was not constant over long timescales (for a summary of his evidence, see Table 6.1). For the oldest eclipses, Dunthorne had found that his tables put the moon at higher longitudes than observed, but for the two solar eclipses observed by Ibn Yūnus the opposite was true: the tables put the moon at lower longitudes than were observed. The eclipses seen by Regiomontanus, Walther, and al-Battānī agreed with the oldest observations in indicating that tables put the moon at too high longitudes. The evidence therefore appeared contradictory, and so Dunthorne was forced to weigh the relative merits of the different sets of data. He chose to put his trust in the observations by Ibn Yūnus since they were observed from a place whose longitude was believed to be known accurately and the observations themselves were more precise and provided consistent results (as much as can be said for only two observations).

Dunthorne concluded by estimating the size of the moon's secular acceleration:

> If we take this Acceleration to be uniform, as the Observations whereupon it is grounded are not sufficient to prove the Contrary, the Aggregate of it will be as the Square of the Time: And if we suppose it to be 10″ in 100 Years, and that the Tables truly represent the Moon's Place about A. D. 700. it will best agree with the before-mentioned Observations.[72]

Finally, Dunthorne gave a table of the correction that needed to be applied to his tables to take into account the moon's acceleration (Fig. 6.3). As Dunthorne explained, the secular equation increases with the square of time. As the moon is accelerating, this means that the moon's motion has been slower in the past and so the longitude of the moon will be in advance of what it would be assuming constant mean motion. In the future, the moon will be moving faster than today and so the moon's longitude will also be greater than it would be on the assumption of constant mean motion. The correction to calculated lunar positions to obtain the real position is therefore a positive correction which increases quadratically with time. Dunthorne's "Error of Tab." is the difference between the true and the tabular longitude in the sense table minus true. The (mostly negative) corrections given by Dunthorne must therefore be subtracted from the longitude calculated by his tables to give the true longitude.

---

[71] Dunthorne, "Concerning the Acceleration of the Moon", p. 171.

[72] Dunthorne, "Concerning the Acceleration of the Moon", p. 171.

Fig. 6.3 Dunthorne's table of the corrections needed to his lunar tables to take into account the moon's secular acceleration (Dunthorne, "Concerning the Acceleration of the Moon", p. 172)

Years before Chrift.	Error of Tab.	Years of Chrift.	Error of Tab.	Years of Chrift.	Error of Tab.
700	$-56'\ 6''$	200	$-12'\ 30''$	1100	$+4'\ 0''$
600	$-49'\ 50''$	300	$-9'\ 20''$	1200	$+4'\ 10''$
500	$-44'\ 0''$	400	$-6'\ 30''$	1300	$+4'\ 0''$
400	$-38'\ 30''$	500	$-4'\ 0''$	1400	$+3'\ 30''$
300	$-33'\ 20''$	600	$-1'\ 50''$	1500	$+2'\ 40''$
200	$-28'\ 30''$	700	$0'\ 0''$	1600	$+1'\ 30''$
100	$-24'\ 0''$	800	$+1'\ 30''$	1700	$0'\ 0''$
A.D.0	$-19'\ 50''$	900	$+2'\ 40''$		
100	$-16'\ 0''$	1000	$+3'\ 30''$		

Dunthorne's table of corrections has a peculiarity, however. Although his values all agree with a quadratic function with a coefficient of $10''$ per century[2] for the quadratic term, in order to obtain agreement with the values Dunthorne derived from the eclipses observed by Ibn Yūnus, Dunthorne's function has a positive offset. If the mean motion of the moon given in the tables is correct for the epoch of the tables, as Dunthorne demonstrated in his 1746/7 paper, then a constant secular acceleration will have the same effect of making all calculated longitudes before the epoch of the tables less than their true positions: the difference between the true and the tabular longitude cannot have been positive at some dates and negative at other dates.[73] In fact, however, Newton's value for the moon's mean motion in his *Theory of the Moon's Motion*, which Dunthorne had used in constructing his tables, is significantly too low for the eighteenth century because it had, like most other estimates of the mean motion, been deduced from comparison of ancient and modern data, and so was appropriate for the mid-point between antiquity and the eighteenth century, roughly the sixth or seventh century A.D. Dunthorne, however, believed that his tables were correct for the eighteenth century—in fact errors in the equations of anomaly must have been compensating for the too small mean motion of the moon. The error in the mean motion, however, dominated on long timescales, and this is why Dunthorne was forced to assume a sometimes positive and sometimes negative secular equation.[74]

Dunthorne's Use of Historical Evidence

In his paper on the secular acceleration Dunthorne developed a new approach to using historical astronomical observations in solving problems within current astronomy. Before Dunthorne, historical observations were either considered too untrustworthy to be of use or were used as if they were contemporary observations with no consideration of their reliability or accuracy. Dunthorne, however,

---

[73]Contrary to Mercier (1994), p. 198, this is not simply a mathematical change in origin.

[74]In addition, recent studies have shown that there was a significant variation to the long-term trend in the rate of rotation of the Earth from about A.D. 500 to 1000; see Stephenson (1997), p. 514. This may also have contributed to Dunthorne's values of the secular equation deduced from the Arabic observations being too great.

pioneered a new approach in which aspects of the observations were used which were not the data the original observer had been concerned with. For example, it had been known at least since Bullialdus and Streete in the mid-seventeenth century that the times of the eclipses reported by Ptolemy were inaccurate, inconsistent, and often ambiguous as to what phase of an eclipse a timing related to. Ptolemy had used the times of these eclipses to derive the parameters of his lunar theory. Dunthorne, however, largely ignored the times of the eclipses and instead used facts such as that the moon set eclipsed, or rose before the eclipse began, to place constraints upon the size of the moon's acceleration. Dunthorne's approach avoided issues such as the inaccuracy of timing methods used by the ancient observers, possible errors in the recording of time measurements or the conversion between one time unit and another, ambiguities in the record, or any adjustment of the time by Ptolemy. This lateral thinking by Dunthorne opened up a source of data that had previously been considered too problematical for studying the moon's long-term motion.

Dunthorne discussed his techniques for obtaining data for studying the moon's secular acceleration from the historical records, and described and analysed the records themselves, but, perhaps surprisingly, he said nothing about the history of astronomy in his 1749 paper. He made no remarks about any of the astronomers whose observations he used, not even the ubiquitous comments on Tycho as expert observer or Ptolemy as the greatest astronomer of antiquity that are found throughout seventeenth- and eighteenth-century astronomical literature.[75] No references are made to any of the several histories of science that had been written over last century. It appears as if for Dunthorne when writing this paper, the records spoke for themselves. What was important was obtaining whatever scientific data that could be squeezed out of the records, not the reputations of the observers. This pragmatic approach was indeed the correct one: it was irrelevant whether Ptolemy or the Babylonians were good observers, when because of his innovative use of eclipses at the horizon all Dunthorne needed was the fact that they had seen the event at all.

Some further insights into Dunthorne's attitude towards and use of historical data can be found by looking at his other work. As discussed above, on 20 June 1738, Dunthorne wrote to Rouse that he had examined al-Battānī's observations of the greatest and least solar meridian passages in order to investigate the change in the obliquity of the ecliptic and the latitude of al-Raqqa, and also compared al-Battānī's observations of star positions with the positions of the same stars in Flamsteed's catalogue in order to derive a value for precession. Dunthorne referred to al-Battānī's observational technique ("with a Parallactic Instrument of Ptolemy, whose side and Alhidase or sight ruler were very long, he carefully observed (for several years)"[76]), but did not comment on the likely accuracy of the observations apparently instead accepting the common opinion that al-Battānī was an accurate observer.

Dunthorne was more critical in his use of historical data in a study of a medieval manuscript containing observations of comets published in the *Philosophical Transactions* in 1751.[77] Dunthorne had found a Latin manuscript in the library of Pembroke Hall containing five (predominantly astrological) tracts concerning comets. From three of these tracts, Dunthorne extracted descriptions of observations of comets seen in 1264, 1301, and 1106. Observations of the comets of 1264 and 1106 were already known from other sources and had been collected by Hevelius in his *Cometographia*. But from the Pembroke Hall manuscript, Dunthorne was able to deduce the orbital elements of the comet and propose from the close agreement of these elements with those deduced by Halley for the comet of 1556 that they might be the same body. However, Dunthorne showed admiral caution, noting that some of the

---

[75]Dunthorne's silence on historical matters in his 1749 paper may be contrasted with comments found in his letters to Rouse. For example, Dunthorne referred to al-Battānī as "that celebrated Astronomer of Antiquity" (MS Oxford 1, p. 77) and Theon of Alexandria as "Theon Alexandrinus Junior father of the Celebrated Hypatia" (MS Museum 95, p. 100).

[76]MS Museum 95, p. 77.

[77]Dunthorne, "Concerning Comets".

orbital elements are "indeed but tentative",[78] and claiming only that "the resemblance of all the elements gives some ground for conjecture, that this comet might possibly be the same"[79] as that of 1556.

The description of the comet of 1301 was "too imperfect for us to attempt determining the orbit therefrom",[80] but Dunthorne justified bringing it to attention in case it was seen again. Interestingly, in discussing the record of this comet in the manuscript, Dunthorne noted that the number 26 given for the latitude of the comet "is a different writing from the rest of the manuscript, and has manifestly been alter'd since it was first written; it seems to have been 16° at the first, which I think the truer reading".[81] This demonstrates the care with which Dunthorne read this manuscript. Similar care in analysing the text is found in Dunthorne's discussion of the third comet, see in 1106. Here Dunthorne used his knowledge of the Muslim and Julian calendars to correct the date of the observation:

> The word Junii here found seems to have been transcribed by mistake for the Arabic month Jumedi.j, the last day whereof that year was Wednesday Feb. 7. A.C. 1106; whereas the last day of June fell upon Saturday. This reading agrees with the following notes concerning the same comet collected by Hevelius in his *Cometographia*, p. 821.[82]

Halley had previously suggested that the comet of 1106 was the same as that which had been seen in 1680. Dunthorne, however, showed that this could not be true: "The wide disagreement there is between the manuscript account of this comet, and its places here computed, must very much lessen, if it does not quite overbalance, the force of the arguments brought by Dr. Halley to prove the identity of these two comets".[83]

Comparing Dunthorne's attitude towards historical data in his 1739 letter to Rouse, his 1749 paper on the moon's acceleration and his 1751 paper on comets it is possible to trace the development of an increasingly critical and rigorous approach to reading historical sources. In his 1739 letter, Dunthorne simply takes the data as fact, but in his 1749 and 1751 papers, he reads his sources carefully, and corrects them on the basis of internal inconsistencies (changing equinoctial hours to seasonal hours in Theon's eclipse report and correcting the date of one of the comet observations in the Pembroke Hall manuscript) and study of the physical appearance of a manuscript. A word of caution must be raised here, however. With only three examples of Dunthorne's use of historical data, it may be misleading to talk of the development of his approach—it may simply be that Dunthorne used whatever approach seemed best for the particular problem at hand.

Two further sources of evidence may provide further information on Dunthorne's knowledge of the history of science. The first is a letter written by Dunthorne to Rouse on 29 April 1764. Unfortunately, most of the letter has been ripped out of Rouse's bound collection, leaving only the upper left quarter of the first page and the left edge of the second page. Rouse described the contents of the letter as follows: "Mr. Dunthorns Acc^t. of ye Grecian Year &c. Observations of Eclipses to illustrate the same—and ye use of ye Tables in ye Common Prayer Book to finding Easter NS".[84] From what can be read of this letter it is clear that Dunthorne fully understood the various chronological systems of antiquity.

The second source is the final book of Roger Long's *Astronomy* which dealt with the history of astronomy. Long died before the completion of this final book, and requested that Dunthorne complete it. Dunthorne, however, died himself before he could finish writing the book and the text was finally completed by William Wales. An advertisement at the end of the book describes what happened:

> In the year 1764, Dr. Long published the third book of his Astronomy, as a part only of his second volume. Before the time of his death he had finished and printed off the fourth, and a small part of the fifth book, viz. the

---

[78] Dunthorne, "Concerning Comets", p. 283.

[79] Dunthorne, "Concerning Comets", p. 285.

[80] Dunthorne, "Concerning Comets", p. 286.

[81] Dunthorne, "Concerning Comets", p. 286.

[82] Dunthorne, "Concerning Comets", p. 287.

[83] Dunthorne, "Concerning Comets", p. 288.

[84] MS Museum 95, unnumbered contents page.

three first chapters of it, and a small part of the fourth, including p. 654. He had indeed printed off the leaf following, which, it was thought proper, should be cancelled, and the new part begins p. 655. A few hints only were left in the Author's own handwriting towards finishing the remainder, which may, now and then, be perceived in the course of the Work. One instance of this kind occurs p. 663, relating to a conversation which he had held with the old Lord Pembroke. A short time before he died, he had desired Mr. Dunthorne, whom he nominated one of his executors, to finish the Work. Mr. Dunthorne made indeed a rough draught of the remaining part of it; but being much engaged in pubic business, by his being appointed superintendant of the works of the Bedford Level Corporation, his avocations became great and unavoidable, and he left it as last in a very imperfect state. After his death, Mr. Wales, F. R. S. Master of the Royal Mathematical School in Christ-Hospital, and Editor of the Original Observations, made by himself and others in the course of a voyage around the world, was prevailed upon to revise and correct Mr. Dunthorne's continuation, and to complete it upon the original plan, sketched out and begun by the Author himself, in that part of the fifth book which had been printed before he died: and it is owing to the care of Mr. Wales that it is now presented to the Public in its present state.[85]

Unfortunately, this advertisement does not make it clear which parts of Long's book are due to Dunthorne. Long apparently wrote the first three chapters—an introduction reviewing previous works on the history of astronomy, a chapter on antediluvian astronomy, and a chapter on the astronomy of "fabulous times"—and the first page of the fourth chapter on Chaldean and Egyptian astronomy. Dunthorne then prepared a "rough draught of the remaining part" which was taken up by Wales who took it upon himself to "revise and correct Mr. Dunthorn's continuation, and to complete it upon the original plan, sketched out and begun by the Author himself". It is therefore uncertain how much of Dunthorne's draft was retained in the final version. Wales's return to Long's "original plan" suggests that Dunthorne's work was at very least significantly rearranged.

Long's own text ended on page 654, four paragraphs into his chapter on Babylon and Egypt. These four paragraphs focused on various fabulous legends about Babylonian astronomy found in classical sources. With page 655, written by Dunthorne or Wales, the text switched to a fairly technical discussion of Babylonian values for the length of the synodic and anomalistic month and the Saros and continued with a description of the Babylonian observations preserved in Ptolemy's *Almagest*, and throughout the rest of the work a fairly technical summary of the history of astronomy is given—more technical and comprehensive, for example, than the accounts given by Flamsteed, Estève, or Weidler. As we do not know how much of the final text is due to Dunthorne we cannot make firm conclusions about Dunthorne's understanding of history of astronomy based upon this work. Nevertheless, it seems likely that, at least late in his life, Dunthorne had a fairly thorough knowledge of the subject, and had read the work of other historians of astronomy quite widely.

Dunthorne's use of historical astronomical data can perhaps be characterised by three things: (1) the careful study of historical sources and, where necessary, their correction based upon internal evidence; (2) lateral thinking in developing methods to exploit historical data that avoids problems of their accuracy or ambiguity; and (3) an avoidance of stating unnecessary historical information. The first two of these were truly innovative developments in the study of historical data.

---

[85]Long, *Astronomy*, unpaginated advertisement.

Chapter 7
An Integrated Approach: Tobias Mayer

Although any one of these things would have been sufficient to commend the man's memory to posterity, he could have done without them all; being knowledgeable about the stars [and] confident that his name would endure as long as the moon. I do not say "moon" at random, gentlemen, since she seems to have revealed her more hidden parts to Mayer as if to a new Endymion.

—Abraham Gotthelf Kaestner, *Elogium Tobiae Mayeri*,
13 March 1762; translation by Forbes (1980), p. 14.

Three years after Dunthorne published the first detailed study of the moon's secular acceleration, Tobias Mayer produced the first set of lunar tables which incorporated the secular acceleration directly into the calculation of the moon's longitude. Dunthorne had provided a table that allowed a correction to be applied to lunar positions calculated from his lunar tables. Dunthorne's correction, however, was applied after the moon's position had been calculated. By contrast, Mayer integrated the secular acceleration within the initial calculation of the moon's mean position. This meant that the effect of the secular acceleration was taken into account when determining the various equations of anomaly (which depend upon the elongation of the mean moon from the sun and the mean moon from the apogee). The difference between applying the correction for the secular acceleration before or after calculating the anomaly is small but not trivial, and Mayer's method was the theoretically correct one. Mayer adopted a value for the size of the secular acceleration that was considerably smaller than that found by Dunthorne. Mayer did not explain how he had derived this value in the introduction to his published tables, but it is possible to reconstruct his general method from his preserved manuscript notes.

Tobias Mayer

Tobias Mayer was born on 17 February 1723 in the town of Marbach in the German state of Württemberg.[1] Following the early death of his parents, Mayer was educated at the Latin school in Esslingen and learnt mathematics in the company of a local shoemaker who had a large collection of

---

[1] The following is based upon the detailed biography by Forbes (1980). See also Forbes (1967) and Wepster (2010), pp. 27–42.

J.M. Steele, *Ancient Astronomical Observations and the Study of the Moon's Motion (1691–1757)*,
Sources and Studies in the History of Mathematics and Physical Sciences,
DOI 10.1007/978-1-4614-2149-8_7, © Springer Science+Business Media, LLC 2012

mathematical books. At the age of 18, Mayer wrote his first scientific treatise, a proposal for a new method for inscribing polygons within circles. In 1744 Mayer moved to Augsburg to take up employment with the publisher Johann Andreas Pfeffel. The following year, Pfeffel published Mayer's *Mathematischer Atlas*, a collection of 60 illustrated plates with marginal notes covering a broad range of mathematical topics: arithmetic, geometry, trigonometry, astronomy, geography, chronology, gnomonics, fortification, artillery, civil architecture, optics, and mechanics.[2] In addition to highlighting Mayer's mathematical and artistic skills, this work also demonstrates Mayer's wide reading of the works of contemporary scientists.

Mayer left Augsberg in 1746 to move to Nuremberg where he worked for the Homann firm of mapmakers. Mayer contributed several maps to the *Gesellschafts Atlas* and undertook a comparison of three different maps of Germany to illustrate the large discrepancies between the latitudes and longitudes of cities found on these maps. Mayer saw an urgent need to improve knowledge of the geographical coordinates of cities through astronomical means; as argued by Forbes, most of Mayer's subsequent astronomical work can be understood as motivated by the goal of improving terrestrial geography.[3]

In 1750 Mayer was offered and accepted a professorship at the Georg-August Academy in Göttingen to begin in the summer of 1751. Over the next 10 years Mayer taught a wide variety of mathematical subjects including mechanics, geometry, algebra, military architecture, mathematical geography, and astronomy.[4] In addition to his lecturing duties, Mayer was jointly responsible with Johann Andreas Segner for running the new observatory built in the town. Segner and Mayer fought bitterly over the observatory until Mayer succeeded in having Segner removed from his position in 1754.

Mayer's work on the lunar theory began during his last few years in Nuremberg and occupied much of his time in Göttingen until the mid-1750s. His first published set of tables appeared in 1753 with the title "Novae Tabulae Motuum Solis et Lunae" in the *Commentarii Societatis Regiae Scientiarum Gottingensis*. Mayer had already gone through several iterations in the development of his lunar tables. In early 1751, Mayer wrote to Delisle that he had produced new tables that were accurate to within 2′,[5] and over the next few years, Mayer drafted several more sets of tables, some of which are preserved in manuscript form. He corresponded extensively with Leonard Euler over the period from 1751 to 1755, describing several further sets of tables which are no longer extant.[6]

Mayer's 1753 tables were generally well received. Euler in particular was enthusiastic about them:

> In coming now to your new important discoveries, I first of all congratulate you wholeheartedly upon them, and wish that their importance would soon be known to everyone. Your first tables have indeed aroused as much applause as amazement, only jealousy has already let itself be seen more than clearly.[7]

Euler was referring to the criticism of d'Alembert, who claimed that Mayer's tables had no basis in theory. Euler hoped that Mayer would refute d'Alembert's arguments. He continued:

> In England one seems to have been fairer since, following the representations which I made on the importance of your tables, I have received such an answer from which I can with reason conclude that you would be regarded as a worthy competitor for the award established for the discovery of longitude. If you could now still add a method of knowing how to determine at sea the position of the Moon through its distance from fixed stars so accurately that the complete longitude [obtained] from the comparison of it with your tables does not vary by more than half a degree from the truth, you could be assured of the prize of 20,000 pounds sterling.[8]

---

[2] Forbes (1980), p. 36.

[3] Forbes (1980), pp. 42–43.

[4] Forbes (1980), pp. 106–133.

[5] Mayer to Delisle 14 January 1751; see Forbes (1983), no. 9.

[6] Forbes (1971).

[7] Euler to Mayer 11 June 1754; translation by Forbes (1971), p. 86.

[8] Euler to Mayer 11 June 1754; translation by Forbes (1971), p. 86.

Mayer followed Euler's advice and submitted his tables to the Board of Longitude. The tables were tested by Bradley, then Astronomer Royal, and their accuracy confirmed, but concerns over the practicability of measuring the moon's position onboard ship, and the advent of the Seven Years' War, meant that the prize was not awarded until after Mayer's death in 1762. His widow was eventually given a partial award of £3,000 in recognition of his work. Shortly before his death, Mayer sent revised and improved lunar tables to the Board from which Nevil Maskelyne extracted the parameters and published revised tables in 1770, which were used for producing the Nautical Almanac. Maskelyne also published Mayer's theoretical treatment of the lunar theory in 1767.

Mayer remained in Göttingen until his death in 1762. In the mid-1750s, Euler arranged an offer of a position at the Berlin Academy for Mayer, but Mayer's resignation of his position in Göttingen was refused. He suffered from ill health in the last years of his life, although he continued his astronomical and other scientific work, publishing many astronomical observations and compiling a new catalogue of zodiacal stars.

Mayer's Novae Tabulae Motuum Solis et Lunae

Mayer's lunar and solar tables published in 1753 in the second volume of the *Commentarii Societatis Regiae Scientiarum Gottingensis* were the first set of tables to include a table for the secular acceleration of the moon. The tables, which I shall henceforth call his 1753 tables,[9] were the only complete set of tables published in Mayer's lifetime, although several other sets of tables are preserved in manuscript form or are described in Mayer's correspondence with Euler and others.[10] Around the end of 1754, Mayer sent revised tables to England for consideration by the Board of Longitude, and following his death a further revised set of tables were sent by his widow to the Board. Nevil Maskelyne later published the coefficients of the first set of these (calling them Mayer's "first manuscript tables") and the full tables of the second (Mayer's "last manuscript tables").[11]

The publication of Mayer's 1753 tables might suggest that these represented a significant step in the progress of his lunar theory. However, it is more appropriate to understand them instead as a snapshot of Mayer's continual development of his lunar tables over the 1750s. During this period Mayer was repeatedly tinkering with the tables: refining their parameters, sometimes slightly altering their layout, and testing their accuracy against observation. He generated more than 30 sets of coefficients for tables, at least 10 of which were worked out (sometimes only partially) into tables.

Mayer prefaced the publication of his 1753 tables with an explanation of the different inequalities in the moon's motion that he has considered, a discussion of the basis on which he has constructed the tables, and a claim for the tables' accuracy. Mayer explained that he has tables for 13 inequalities to be used in calculating the moon's longitude. Three of these are the most important: XI, the equation of centre; XII, the evection; and XIII, the variation. The remaining ten equations are smaller and depend upon these three. Mayer said that he deduced the inequalities in the moon's motion from the theory of Newton which Euler had formulated into general analytical equations. However, it would take too long, he said, to explain how he had solved these equations and so he will merely describe their character. In addition to not describing the theoretical basis for his tables, Mayer's preface gave no explanation for his choice of parameters. The only exception to this is the value of the mean motion of the moon, which Mayer took to be somewhat greater than was found in the tables of other astronomers.

---

[9] Wepster (2010) denotes these tables by the code "kil".

[10] A list of Mayer's lunar tables (including cases of known table parameters and equations which may not have eventually been used by Mayer to draw up tables) is given by Wepster (2010), p. 212.

[11] Wepster (2010), pp. 210–211.

An understanding of the development of Mayer's lunar tables has only recently been achieved through the work of Wepster.[12] An important aspect of Mayer's approach was the refinement of the coefficients of the 13 equations of anomaly through comparison with observations of the moon's position. Mayer gathered a large collection of timed observations of the moon's position, timed lunar and solar eclipse observations, and timed occultations of stars by the moon. These included his own observations, those sent to Mayer by other astronomers, and others that he had gathered from published sources. In a letter to Euler dated 7 May 1573, Mayer reported that he also had

> a total of 139 observations of the Moon, dating from September 1743 up to April 1745, have also been put into my hands by a man called Schumacher, who travelled through here a few months ago and who may be not unknown to you. … I have compared with the tables most solar and lunar eclipses which were observed after the beginning of the last century—particularly since the invention of telescopes and pendulum-clocks … those longitudes of the ☽ which Dunthorne quoted in the [Philosophical] Transactions out of the observations of Flamsteed[13]

The observations brought by Christian Schumacher were made by Bradley at the Greenwich observatory, and were sent by Bradley's assistant Gael Morris to Euler, who in turn asked Schumacher to pass them on to Mayer.[14] Mayer's collection of seventeenth century eclipse records is preserved in the manuscript MS Mayer 15_{48}, a slim bound quarto entitled "Historia Eclipsium". In this manuscript Mayer catalogued the details of historical eclipses covering the period from 1610 to 1678. For each eclipse, Mayer noted the source of the record (Curtius's *Historia Coelestis*, Riccioli's *Astronomiæ Reformatæ*, Bullialdus's *Astronomia Philolaica*, or Le Monnier's *Histoire Celeste*), details of the observation, and a list of the dates of eclipses separated by intervals of one Saros from the date of the observation to Mayer's time. Mayer marked the dates of some of the eclipses in these lists with a dot to indicate that he had a record of the observation of the eclipse on that date. The latest eclipse date in these Saros lists is 10 November 1761 (f. 5v) but the latest date indicated with a dot is 25 February 1747 (f. 7r), which seems to suggest that Mayer produced this compilation in 1747, around the time when he first started work on lunar theory.

Mayer's method for refining his lunar tables can be summarized as follows.[15] Starting with a set of lunar tables, Mayer calculated the position of the moon at the moment of an observation of the moon's position and recorded the difference in the computed and observed lunar longitude. More than 350 examples of such calculations are found in the manuscript MS Mayer 15_{41} alone. Mayer then gathered the results of his calculations into large tables; Wepster has named these large tables "spreadsheets" because of their similarity in appearance and to a certain extent use to modern spreadsheets and I will use this term to avoid confusion between these spreadsheet tables and lunar tables. Several spreadsheets are preserved in the manuscript MS Mayer 15_{33}, along with a few inserted among the calculations of lunar positions and notes on lunar theory in MS Mayer 15_{28} and MS Mayer 15_{41}. In the left-hand column of a spreadsheet, Mayer wrote the date of a lunar position observation. In the next column Mayer gave the error in the calculation of the moon's position at the moment of the observation. Subsequent columns give the error if the coefficient of one of the equations of anomaly is altered by a stated amount (for example $VI = 1 - 2/16$, meaning that the coefficient of equation VI is reduced by 2/16ths). From these spreadsheets, Mayer was able to deduce the optimal group of corrections to reduce the error in the calculated positions to as close to zero as possible. He then produced an improved set of lunar tables with these revised coefficients. It is not known how Mayer was able to identify from

---

[12] Wepster (2010).

[13] Forbes (1971), p. 65.

[14] Euler explains the origin of the observations given to Mayer by Schumacher in a letter dated 15 May 1753. See Forbes (1971), p. 68.

[15] The following discussion is based upon the work of Wepster (2010).

the mass of data in the spreadsheets which combination of corrections to the coefficients to accept, but it seems that on most occasions his selections did indeed result in an improvement to the tables.

Mayer went through this process of refining the coefficients of the equations of anomaly in his lunar tables several times during the 1750s. The 1753 tables were simply a step in this iterative process which Mayer judged had reached a level of accuracy that made them worth publishing. Mayer wrote to Euler that his tables agreed with almost all the observations available to him to within 1′, with the exceptions disagreeing by only up to about 1′30″.[16] Nevertheless, straightaway Mayer returned to his spreadsheets to refine the tables further and continued making improvements to the tables until at least 1757.

The accuracy of Mayer's 1753 tables, which was a significant improvement over any existing tables, was applauded by Euler who wrote to Mayer on 26 February 1754:

> Your *Tabulae Lunares* cannot be otherwise regarded as the most admirable masterpiece in theoretical astronomy, and I should never have guessed that by this means the tables would have been brought to such a degree of completeness. As I have been so little fortunate in this task, I therefore value it all the more highly, because I first of all placed the lunar tables on the footing that they were plagued before everything else by the variable eccentricity and the astonishing motion of the apogee, which things would have been impossible to harmonise with the true theory. Now you have done everything that from a practical point of view can ever be desired; only in the abstract theory do I see much incompleteness.[17]

Euler enthusiastically promoted Mayer's tables and encouraged him to submit them to the Board of Longitude in England. In August of the following year a letter appeared in *The Gentleman's Magazine* describing the tables which

> having been highly recommended by a very celebrated professor abroad; and some of our most eminent *English* astronomers most approving of their form, as I have heard, I here send you the author's own account of them; by inserting which in your magazine, you will doubtless oblige all lovers of astronomy, and navigation.[18]

The letter, signed "B. J.", was written by John Bevis,[19] a well-connected doctor and astronomer. In it, Bevis paraphrased the preface to Mayer's 1753 tables (the text is an almost complete rendering in English of Mayer's Latin text, with a few sentences compressed or omitted). At the end of the letter is appended a short remark about the reports of the Ibn Yūnus eclipses in Curtius's *Historia Coelestis*, and a critical comment about the accuracy of Mayer's tables:

> And furthermore, notwithstanding the extraordinary assurances of the author, it has been found on calculating only 6 or 7 observations of the moon, taken by Dr. Halley at Greenwich, (the computations being repeated by different persons,) that they err considerably about 2 min in one, and no less than 4′37″ in another, to wit, that of 30 March 1726.[20]

It is not certain whether these remarks were written by Bevis or appended to the letter by the editor of *The Gentleman's Magazine*. In the next issue, however, a correction appeared reporting that consultation of Halley's manuscripts had revealed that the 4′37″ error for the moon's position on 30 March 1726 was due to a mistake in the published edition of Halley's observations and "when rightly reduced is well enough represented by Mr *Mayer's Tables*".[21]

---

[16] Mayer to Euler 7 May 1753. See Forbes (1971), pp. 65–66.

[17] Euler to Mayer 26 February 1754; translation by Forbes (1971), p. 79.

[18] Bevis, "Mayer's new Tables of the Sun and Moon".

[19] Forbes (1980), p. 142.

[20] Bevis, "Mayer's new Tables of the Sun and Moon", p. 376.

[21] *The Gentleman's Magazine* 24 (September 1754), p. 439.

Fig. 7.1 Mayer's table for the secular equation in his 1753 lunar tables (Mayer, "Novae Tabulae Motuum Solis et Lunae", table XXV) (Courtesy Niedersächische Staats- und Universitäts-Bibliothek, Göttingen, shelfmark 8 PHYS MATH IV, 330)

(XXV.)

Acceleratio motus medii lunae, addenda semper ad longitud. mediam ☽		Pro loco Nodi & Anomalia media Lunae. *Argum. Anomalia media Solis.* Simpla haec aequatio pro longitudine Nodi ascendentis, Duplum eius pro Anomalia media Lunae.						
anni	0 ′ ″	0 +	1 +	2 +	3 +	4 +	5 +	
		′ ″	′ ″	′ ″	′ ″	′ ″	′ ″	
800	I. 9. 48	0 0. 0	5. 1	8. 47	10. 18	9. 4	5. 17	30
700	I. 4. 19	1 0. 11	5. 10	8. 53	10. 18	8. 59	5. 8	29
600	0. 59. 4	2 0. 22	5. 19	8. 58	10 18	8. 53	4. 58	28
500	0. 54. 3	3 0. 32	5. 28	9. 3	10. 18	8. 47	4. 48	27
400	0. 49. 15	4 0. 43	5. 37	9. 9	10 17	8. 41	4. 38	26
300	0. 44. 40	5 0. 53	5. 46	9. 14	10. 17	8. 35	4. 28	25
200	0. 40. 19	6 1. 4	5. 55	9. 19	10. 16	8. 29	4. 18	24
100	0. 36. 11	7 1. 14	6. 3	9. 23	10. 15	8. 22	4. 8	23
0	0. 32. 16	8 1. 25	6. 12	9. 27	10. 14	8. 16	3. 58	22
100	0. 38. 35	9 1. 35	6. 20	9. 31	10. 13	8. 10	3. 48	21
200	0. 25. 7	10 1. 45	6. 28	9. 35	10. 12	8. 3	3. 38	20
300	0. 21. 53	11 1. 55	6. 36	9. 39	10. 10	7. 56	3. 27	19
400	0. 18. 52	12 2. 5	6. 44	9. 42	10. 8	7. 48	3. 17	18
500	0. 16. 5	13 2. 15	6. 52	9. 46	10. 6	7. 41	3. 6	17
600	0. 13. 31	14 2. 25	7. 0	9. 50	10. 4	7. 34	2. 56	16
700	0. 11. 10	15 2. 35	7. 8	9. 53	10. 2	7. 26	2. 45	15
800	0. 9. 3	16 2. 45	7. 16	9. 56	9. 59	7. 18	2. 35	14
900	0. 7. 9	17 2. 55	7. 23	9. 58	9. 56	7. 10	2. 24	13
1000	0. 5. 28	18 3. 5	7. 31	10. 1	9. 53	7. 2	2. 13	12
1100	0. 4. 1	19 3. 15	7. 38	10. 3	9. 50	6. 54	2. 2	11
1200	0. 2. 48	20 3. 25	7. 45	10. 5	9. 46	6. 46	1. 51	10
1300	0. 1. 47	21 3. 35	7. 52	10. 7	9. 43	6. 37	1. 40	9
1400	0. 1. 0	22 3. 44	7. 58	10. 9	9. 39	6. 29	1. 29	8
1500	0. 0. 27	23 3. 54	8. 5	10. 11	9. 35	6. 20	1. 18	7
1600	0. 0. 7	24 4. 4	8. 12	10. 13	9. 31	6. 12	1. 7	6
1650	0. 0. 2	25 4. 14	8. 18	10. 14	9. 27	6. 3	0. 56	5
1700	0. 0. 0	26 4. 23	8. 24	10. 15	9. 23	5. 54	0. 45	4
1750	0. 0. 2	27 4. 33	8. 30	10. 16	9. 18	5. 45	0. 34	3
1800	0. 0. 7	28 4. 42	8. 36	10. 17	9. 14	5. 36	0. 23	2
1850	0. 0. 15	29 4. 52	8. 42	10. 18	9. 9	5. 27	0. 12	1
1900	0. 0. 27	30 5. 1	8. 47	10. 18	9. 4	5. 17	0. 0	0
1950	0. 0. 42							
2000	0. 1. 0		11	10	9	8	7	6

Com. Soc. Gott. Tom. II.   Ggg

Mayer's Analysis of the Secular Acceleration of the Moon

Table XXV of Mayer's 1753 tables gives the correction to the mean longitude of the moon to take into account the secular acceleration in the moon's mean motion (Fig. 7.1). The table gives this correction to the longitude in degrees, minutes, and seconds for every century from 800 B.C. to

A.D. 1600 and every 50 years from A.D. 1600 to A.D. 2000.[22] This correction is to be applied to the calculation of the moon's mean position after a preliminary value has been determined from the year, month, day, and hour using tables XV to XXIV, to give the final value of the moon's mean position that is to be used in calculating the corrections for the anomalies.

The values of the correction for the secular acceleration given in Mayer's table are calculated using a quadratic function of centuries from A.D. 1700 with constant 6.7″ per century[2]. In the introduction to the tables, Mayer explained that he had discovered that the moon's mean motion was sensibly more in his day than it had been in the past, clear evidence that the moon's motion has accelerated, although he stopped short of providing a full explanation of how he derived the size of the acceleration. Nevertheless, Mayer's discussion is worth quoting in full as it was his only published words concerning the moon's secular acceleration:

I have spared no labour in establishing the moon's mean motion, as surely as the observations of ancient time allow. I examined, therefore, the most ancient Babylonian, and likewise Hipparchus's and Ptolemy's, observations of lunar eclipses; although these are so coarse that it is vain to attempt to represent them with even moderate agreement; nor will this be seen to be strange to anyone, who considers, that the ancients noted the times of this kind of phenomena to a third or a half of an hour without much care. Furthermore, there is no slight suspicion, that Ptolemy, from whom we learn of these eclipses, changed the times of some of them too boldly, to accommodate the numbers of his own hypothesis. Of which thing indications have been brought forth by ISM. BULLIALDVS in Astr. Philol. L. III, C. VIII. Let no man therefore object if he finds my tables to deviate in one or another of these eclipses by more than half an hour.

Yet, notwithstanding either the carelessness of the ancients or Ptolemy's insincerity, these observations unanimously demonstrate that the motion of the moon was formerly sensibly slower than is found in our age. The same acceleration in the motion of the moon has been taken notice of by Halley, and a few others, but how large it is has hardly been well determined. For this reason, to define it exactly, with much industry, I have surveyed the intermediate observations between Ptolemy and our own; namely Albategnius and other Arab astronomers. Among these I have found two solar eclipses, that on account of their singular circumstances, namely observations of the altitude of the sun at the beginning and end, uniquely among the ancient ones, for which the time can be safely established; and which in the astronomy of the moon are therefore to be held, in my opinion of course, more valuable than gold or silver. Yet I do not remember a single one of those who have complied tables of the motion of the moon who have used the observations of them, though perhaps any one of these of the sun would have been more profitable than all those of Ptolemy. On account of which, and especially because they are extremely useful in demonstrating the acceleration of the moon, they appear to me worthy of being transcribed from Prolegom. Hist. coelestis Tychonis, where they appear among other hitherto obscure texts.

"Anno Hegirae 367 die Iouis, qui erat 28 Rabie posterioris observatum fuit Cahirae in Aegypti metrolopi initium eclipsis solaris, cum altitudo solis effet 15°43′, quantitas obscurationis 8 digit. Ea finita sol elevabatur 33½ gr.

"Anno eoden, die Sabbathi, videlicet 29 mensis Sywal eclipses solis occupavit digitos 7½; in principio sol altus fese 56°, in fine sol occiduus elevabatur gradibus 26."

Another eclipse is quoted at the same place, but of the moon, which for that reason I have willingly passed over.

"Hae tres observations habitae sun tab *Ibn-Iunis*, qui jussu Regis *Abu-Haly Almansor* sapientis Aegypto tunc imperantis rebus vacabat coelestibus. Hujus auctoris tabulas habet *Iac. Golius* Professor Lugd. (qui mihi inde communicavit istan), in quibus plures aliae sui & superirois aevi observationes extant. Locus observationis propinquus urbi Cahiro"*

[footnote: * When this had already been handed to the press, I saw in the Transact. Phil. N. 492 the most celebrated R. Dunthorne brought forward the same eclipses in the same use, viz. to determine the acceleration of the motion of the moon. I do not wish, however, to feel jealous of him when they had barely started to become known among astronomers.]

The date of the former eclipse in Julian years corresponds to A.D. 977 December 13, the beginning we gather from the altitude at 8h24′24″, end 10h43′44″ am, supposing the elevation of the pole at the city of Cairo is 30°2′30″, as determined by recent observations. The other eclipse happened according to our custom on A.D. 978 June 8, the beginning from the altitude of the sun at 2h30′16″, end 4h50′24″ pm. My tables of the moon agree with these and especially the end times within a single minute or two, which will be clear from any trial. But all other tables show them nearly half an hour earlier, an infallible indication that the moon moved more

---

[22] The table as printed contains one typographical error: the correction for A.D. 100 is given as 0°38′35″ instead of 0°28′35″.

quickly now than in times past, and likewise that the quantity of this acceleration, which is shown in these tables, is well defined.

Nor, indeed, is this acceleration of the motion of the moon so small, that it can be clearly be shown from the observations from this century and the last. I have found the mean motion of the moon in 60 years, in our century, $1^{s}10°43'24''$, whereas other tables, whose authors obtained the mean motions from the most ancient compared with the more recent observations, give only $1^{s}10°41'10''$ or at most $1^{s}10°42'15''$. I have considered many observations of eclipses from more than sixty years before our time, taken was equal care and diligence as is usually the custom today. Almost all of these I have computed by the tables with this swifter motion and not found any which argues for an error in longitude greater than one minute; this error, however, would have grown to 3 minutes if I had retained the common quantity for the mean motion. And in a particular way I undertook to set the quantity of this motion most surely, notwithstanding the imperfection of my tables, if by chance there were yet any. I selected eclipses that are at the interval of one Chaldean Period, to wit, 223 lunations, or, which is better, several periods apart; in these intervals are restored practically all the anomalies of the moon, and so the errors in the tables must also as nearly as possible repeat. If therefore it is established how much the tables differ from observations at the beginning of a period, it is also known how much it ought to be after one or more have gone by; and so if unequal errors are found, that is a sign that the mean motion ought to be corrected by a quantity which will bring them back into equality. In this way, not just one or two cycles, but many were subjected to examination, so that I should venture that among the eclipses observed in our and the preceding century very few remained which have escaped examination.

Many other somewhat older lunar observations, Tycho of course, Walther and Regiomontanus, both in syzygy and outside it, I compared these with the tables and everywhere I found so great agreement as can be as expected from those slightly more coarsely observed.[23]

Let us examine Mayer's discussion in more detail. In the first paragraph, Mayer says that in order to determine the moon's mean motion he has examined the ancient observations by the Babylonians, Hipparchus, and Ptolemy, but found that they are so imprecisely recorded and inaccurate that he cannot find a value that satisfies all of them. Furthermore, he thinks that Ptolemy may have altered the times of some of them in order to obtain agreement with his own tables, citing Bullialdus. As a result, the reader should not be surprised if Mayer's tables do not always agree with the observations reported by Ptolemy. The problem of obtaining agreement between lunar tables and the observations in Ptolemy's *Almagest* had, of course, a long history going back to the work of Longomontanus, Bullialdus, and Thomas Streete.

In the next paragraph, Mayer addressed the acceleration of the moon's motion. He says that despite the problems of the ancient observations, they provide clear evidence that the moon's motion was slower in antiquity than it is in his own day. Mayer does not explain how he came to this conclusion and I will return to this issue below. Mayer next remarks that this acceleration has been noted by Halley and others, but that up to now no one has determined the magnitude of the acceleration. It is not certain who Mayer is referring to by the "others"; as I have discussed in Chap. 4, by the end of the 1740s the existence of the secular acceleration of the moon was acknowledged by several astronomers. But it seems that Mayer is not referring to Richard Dunthorne here, however. Dunthorne's determination of the size of the acceleration was published in 1749, but later in Mayer's discussion, when describing the eclipses of Ibn Yūnus, he remarks in a footnote that it was only while his own article was in press that he learnt of Dunthorne's paper. Mayer does not, however, discuss Dunthorne's result (a greater value for the acceleration than Mayer found), and it is interesting that Mayer chose to add the footnote not where he claims to be the first to determine the value of the secular acceleration but instead following his quotation of the Ibn Yūnus eclipse accounts.

Mayer continues this paragraph by discussing the two solar eclipse observations of Ibn Yūnus, which Mayer says are "more valuable than gold or silver". Believing that he is the first person to pay attention to these observations, Mayer quotes Schickard's Latin translation from Curtius's *Historia Coelestis* "because they are extremely useful in demonstrating the acceleration of the moon". Mayer gives his calculations of the times of these two eclipses derived from their observed altitudes and

---

[23] Mayer, "Novae Tabulae Motuum Solis et Lunae", pp. 388–391. In making this translation I have often referred to Bevis's English paraphrase of the passage in his letter to *The Gentleman's Magazine* of 1754.

Table 7.1 Mayer's analysis of values of the moon's mean motion and motion of the apogee for 60 years in various astronomical tables (from Mayer 15$_{41}$, f. 2r)

Tables	Motion of the moon in longitude	Motion of the apogee
Bullialdus	1^{s}10°41′10″	9^{s}11°35′6″
Kepler	1^{s}10°41′19″	9^{s}11°32′34″
de la Hire	1^{s}10°42′1″	9^{s}11°32′34″
Flamsteed	1^{s}10°42′15″	9^{s}11°30′45″
Comparison Flamsteed and al-Battānī	1^{s}10°42′53″	9^{s}11°17′42″
Longomontanus	1^{s}10°41′27″	9^{s}11°34′52″
Cassini	1^{s}10°41′55″	9^{s}11°32′34″

claims that they are "an infallible indication that the moon moved more quickly now than in times past, and likewise that the quantity of this acceleration, which is shown in these tables, is well defined". The impression given here is that Mayer has used these two observations to determine the size of the moon's acceleration. Again, I will return to this issue below.

In the next paragraph Mayer presented further evidence for the existence of the secular acceleration. He explained that from recent observations he has determined that the mean motion of the moon in 60 years is 1^{s}10°43′24″ (over a whole number of revolutions of the ecliptic). However, he says, other lunar tables have significantly lower values for the mean motion in 60 years because their authors have computed the mean motion by comparing ancient observations with their own. Because the moon was moving slower in the past, taking an average over a long timescale will cause the mean motion to be lower than that determined from recent observations alone. Mayer quotes the lowest and highest values for the mean motion in 60 years found in other tables, 1^{s}10°41′10″ and 1^{s}10°42′15″, which are 2′14″ and 1′9″ smaller than his own value respectively. Although he does not say so, the two values he quotes are taken from Bullialdus and Flamsteed. The manuscript MS Mayer 15$_{41}$, f. 2r gives a longer list of values of the mean motion and motion of the apogee in 60 years from the tables of Bullialdus, Kepler, de la Hire, Flamsteed, Longomontanus, and Cassini, plus a comparison of Flamsteed with al-Battānī (Table 7.1). The comparison of Flamsteed with al-Battānī results in a higher value for the mean motion in 60 years than the tables which are based upon averages stretching back to antiquity, in agreement Mayer's argument.

Mayer explains that he has determined his higher value for the mean motion by comparing observations separated by multiples of the 223-month Saros period, after which there will be a close return in the moon's anomaly. Thus, Mayer continues, if the mean motion is correct the error between observations and his tables should be more or less equal for observations separated by Saros intervals, providing a method for determining what the mean motion should be. Mayer says that he has tested almost all of the eclipses of the current and past century to check his value for the mean motion. Finally, he says, he has checked his tables against the observations of Tycho, Walther, and Regiomontanus and found as good agreement as can be expected with their observations.

Mayer's discussion of the moon's mean motion is more detailed than his discussion of the equations of anomaly, but still does not provide a full account of how he obtained his values for the mean motion itself or for the secular equation. He implies, but does not explicitly say, that he has obtained the latter by using the two solar eclipses observed by Ibn Yūnus, and confirmed it by considering the values of the mean motion found in other astronomical tables compared with his own value. But nowhere does he say that this is exactly what he has done, or what method he used (or could be used). Two of Mayer's manuscripts, however, provide some evidence for the way that he obtained his value for the secular acceleration.

MS Mayer 15$_6$ is a small bound quarto containing Mayer's lectures on lunar parallax and various notes and calculations concerning lunar theory. Folios 35–38 contain an analysis of lunar positions for dates in February and March 1749. This analysis includes references to Euler's lunar tables, which Mayer is known to have used from about 1746 to about 1750. Folios 32v–33r relate to the secular acceleration and also refer to Euler's tables. It therefore seems plausible to assume that this manuscript contains material from around 1749 or 1750, early in the development of Mayer's lunar tables. MS Mayer 15_{41} is a much thicker quarto of almost 400 pages bound in three parts. It contains more than 300 calculations of lunar position, various notes on lunar theory, miscellaneous calculations, spreadsheets, and lunar tables. As first recognized by Wepster, the numbering of the folios in the manuscript does not reflect Mayer's original ordering: several of the quires have clearly been bound out of order.[24] The manuscript contains material covering the period from about 1750 to about 1755. Scattered throughout the manuscript are notes and calculations related to the secular acceleration. Unfortunately, because of the confusion in the ordering of the manuscript when it was bound and numbered, it is not always possible to place the notes and calculations on the secular acceleration in the order in which they were written. It seems almost certain, however, that MS Mayer 15_{41} contains material later in date than MS Mayer 15_6.

Neither MS Mayer 15_6 nor MS Mayer 15_{41} provides a full explanation for how Mayer derived a value for the moon's secular acceleration. The two manuscripts contain only working notes on the problem and it is quite possible that Mayer made further notes and calculations which are not pre-served. In what follows I try to present a plausible reconstruction of the development of his under-standing of the size of the acceleration based upon the surviving evidence. The picture that will emerge from this reconstruction is that Mayer obtained his value for the secular acceleration through a gradual process of trial and error, combining evidence from a variety of sources. The proof that his value was correct could come only from the success with which his tables agreed with the observa-tional record. Just as he had been unable to justify theoretically how he had adjusted the equations of anomaly in his tables using observations, he could not explain exactly how he had found his value for the size of the secular acceleration.

The first stage in Mayer's study of the secular acceleration of the moon began around 1749. In MS Mayer 15_6 folio 33r and extending into some empty space at the bottom right of folio 32v, Mayer examined Edmond Halley's analysis of al-Battānī's *zīj* and considered the effect of an acceleration in the moon's motion on the moon's position and mean velocity at various epochs. Mayer was presum-ably already aware of Halley's claim to have identified a secular acceleration of the moon either from Halley's announcement in his *Philosophical Transactions* paper of 1695 or from Newton's words in the second edition of the *Principia*. He was also certainly aware of Kepler's discovery of the secular acceleration and deceleration in the motions of Jupiter and Saturn, and of Euler's claim to have dis-covered a secular acceleration in the motion of the Earth around the sun. Euler's tables, which Mayer had used since 1746, included a secular equation for the sun's position. As I will discuss later in this chapter, Mayer rejected Euler's claims for a secular acceleration of the Earth, but the idea of secular accelerations in the motions of bodies in the solar system was clearly not new to Mayer when he began to consider the possibility of a secular change in the moon's motion.

The top half of folio 33r of MS Mayer 15_6 contains Mayer's analysis of Halley's reconstruction of al-Battānī's lunar tables. Halley had deduced the mean position of the moon for the years A.D. 881, 882, 883, 891, and 901 underlying al-Battānī's tables.[25] Assuming that the longitude difference between al-Raqqa and Paris is 2h25', Mayer adjusted Halley's reconstruction to the meridian of Paris by calculating the mean motion of the moon in 2h25' using Euler's tables and adding the result, 1°20', to the positions given by Halley. Mayer then calculated the difference between the successive posi-tions to deduce the yearly mean motion (see Table 7.2). Mayer was probably hoping to use this information to deduce the value for the mean motion of the moon in al-Battānī's tables, but this data was not precise enough to show any difference in the mean lunar velocity between al-Battānī and his own time. Mayer also computed the moon's mean position for A.D. 901 to compare with the moon's

---

[24] Wepster (2010), p. 151.

[25] Halley's reconstruction is given in his "Emendationes ac Notæ in vetustas *Albatênii* Observationes Astronomicas, cum restitione Tabularum Lunisolarium eiusdem Authoris", p. 920.

Table 7.2 Mayer's analysis of Halley's reconstruction of the mean position of the moon from al-Battānī's tables

Year	Mean longitude	Difference
881	$7^s28°49'$	
882	$0^s8°13'$	$4^s9°24'$
883	$4^s17°36'$	$4^s9°23'$
891	$3^s29°2'$	$11^s11°26'$
901	$0^s12°24'$	$8^s13°22'$

position from al-Battānī's tables, finding a difference of 9′, but does not seem to have pursued this comparison further. Mayer wrote that his computation was "ex tabb. Eul.", but the value he gives is 5′ lower than found in Euler's tables suggesting that Mayer was working at this time with his own revised version of Euler's tables.

Mayer's analysis of the reconstructed tables of al-Battānī was not leading anywhere, so he moved onto another set of data in the bottom half of the page. Mayer listed three values for the mean motion of the moon in 60 years: $1^s10°43'15''$ for A.D. 1700, $1^s10°42'53''$ for A.D. 1325, and $1^s10°41'30''$ for A.D. 600. Mayer did not give the sources for these values. The figure for A.D. 1700 is probably his own value (as I have mentioned, he adopted the slightly higher value $1^s10°43'30''$ in his 1753 tables). The year A.D. 1325 is halfway between A.D. 900 (al-Battānī's time) and A.D. 1750 and the value he gives is the same as that given in MS Mayer 15_{41}, f. 2r for a comparison of al-Battānī and Flamsteed. A.D. 600 is halfway between A.D. 1750 and 550 B.C. (approximately the middle of the Babylonian observations known from Ptolemy); I do not know from where Mayer gets this value for the moon's motion. Mayer next compared the values of the moon's motion in 60 years for A.D. 1325 and A.D. 600 with his baseline figure for A.D. 1700 and found that there is a difference of 22″ between A.D. 1700 and A.D. 1325 and of 1′25″ between A.D. 1700 and A.D. 600. A difference of 22″ in 375 years corresponds to a difference in 1 year of 3‴ 31⁗ which Mayer rounds to 3 5/10‴, and a difference of 1′25″ in 1150 years (Mayer makes a small mistake here as he should take 1100 years) is 4‴ 26⁗ which he rounds to 4 4/10‴. He therefore concluded that the moon's motion in 60 years increases by about 4‴ per year; this means that the moon's motion in one year increases at a rate of 4⁗ per year. Mayer used this value as the basis for several of his subsequent investigations of the moon's secular acceleration.

In some empty space in the bottom right corner of f. 32v, Mayer explored the consequences of the secular acceleration on the moon's velocity and position, correctly showing that as the annual mean motion of the moon increases by 4⁗ every year, the increase in the moon's mean position each year increases by 4⁗/2 multiplied by the square of the year since the 1700 epoch. Mayer returned to this subject in MS Mayer 15_{41} ff. 1r–2v, which was probably written somewhat later as Mayer refers to the higher value of the moon's mean motion found in his 1753 tables. Treating the issue now purely theoretically, he shows that for an acceleration α the lunar velocity increases as αt, where t is the number of years since the epoch, and the longitude will increase according to $\frac{1}{2}\alpha t^2$. Thus, if the moon's secular acceleration is 4⁗ per year, this means that the correction to the moon's longitude for year t is $2⁗t^2$. Converting t into centuries, the correction to the moon's longitude becomes $5''33'''20⁗t^2$, which Mayer rounds to $5\frac{1}{2}''t^2$. It seems that Mayer considered other possibilities for the size of the acceleration as he was writing this discussion. At the top of f. 2v, in a hard-to-read note written half in Latin and half in German, Mayer says that from the "observations of al-Battānī", the acceleration is between 6⁗ and 7 2/10⁗ per year, resulting in a coefficient for the secular equation of between 8 1/3″ and 10″ per century. I do not know the source of these numbers.[26]

---

[26] It is possible that the 6⁗ per year comes from comparing the mean motion over 60 years for al-Battānī's epoch and Mayer's revised value of the mean motion of $1^s10°43'30''$ for A.D. 1700. This would give an acceleration of 5;55⁗ per year which would naturally round to 6⁗ per year.

In this first stage of Mayer's examination of the secular acceleration, Mayer worked exclusively with values for the mean motion of the moon (generally over 60 years), and obtained a preliminary value of the secular acceleration of 4'''' per year (with an alterative value of between 6'''' and 7 2/10'''' per year), from which he derived a coefficient of the secular equation of 5½" per century[2] (or between 8 1/3" and 10" per century[2] for his alternative value of the acceleration). Mayer's next step was to try to refine the size of the acceleration. To do so he tested ancient observations and, perhaps surprisingly, calculations of lunar positions using other set of astronomical tables, against his own calculations. In his own calculations Mayer used a preliminary value of the coefficient of the secular equation. Looking at the resulting error in his calculated position provided Mayer with information on how this preliminary value needed to be modified. Since he was now dealing with observed or calculated longitudes rather than mean motions, in the next part of his study Mayer always considered the coefficient of the secular equation ($c = \frac{1}{2}\alpha$) rather than the acceleration (α) itself.

Mayer began the process of refining the coefficient of the secular equation by analysing the reports of lunar eclipses and occultations of stars by the moon recorded in Ptolemy's *Almagest* and the two solar and one lunar eclipse reported by Schickard from the manuscript of Ibn Yūnus.[27] Figure 7.2 shows a typical example of Mayer's analysis of one of these records: the lunar eclipse of 1 September 720 B.C.[28] According to Ptolemy, this eclipse was seen in Babylon and Ptolemy deduces that the eclipse began 3½ equinoctial hours before midnight. Mayer begins with the date and time of the eclipse. He gives the year as 720/719 meaning 720 B.C. = the Julian year −719. From Ptolemy, he takes the moment of mid-eclipse as 3½ equinoctial hours before midnight, or 8h30' noon epoch. Mayer assumes that the difference in geographical longitude between Babylon and Paris is 2h46' and subtracts this amount to find that the moment of mid-eclipse was a 5h44' at Paris. He then applies a correction of −14' for the equation of time, to arrive at a time for the opposition of sun and moon of 5h30' pm at the meridian of Paris. Next, Mayer calculates the position of the sun and the moon at that moment. His basic calculations are given in the five-column table near the top of the page. From left to right these columns are the longitude of the sun, the sun's apogee, the moon, the moon's apogee, and the moon's node. The first four lines of the table (three rows for the sun's apogee) are the values for the mean motions corresponding to the century, year, day, and hour of the observation. Mayer then adds a correction (58'30") for the secular acceleration of the moon, given in line 6 of column 3, to the moon's mean position to find the final mean position of the moon (11^{s}1°47'20"), in line 7 of column 3. Next, Mayer uses this value of the moon's mean position to calculate the ten equations of anomaly in the secondary table written below and to the right of the main table. This value is added to the mean position, and a further two anomaly corrections are made, to give the true position of the moon (11^{s}0°21'37"). Finally, Mayer adds six signs to the true position of the moon to give the point opposite the moon on the ecliptic and compares this with the true position of the sun given in column 1 (5^{s}1°4'40") and calculates the difference to be −43', which is written below column 3. This final figure is the error in Mayer's tables compared with the observation. The calculations are all made with tables very similar to Mayer's published tables of 1753, except for the coefficient for the secular equation which is calculated using a preliminary value of 6" per century.

Mayer's analysis of the historical observations of eclipses and occultations does not appear to have followed any obvious order. He started with the oldest record from the *Almagest*, the Babylonian eclipse of 19 March 721 B.C., then considered the two solar eclipses of Ibn Yūnus, next the four lunar eclipses observed around Ptolemy's time, then the Timocharis occulations, and finally the remaining *Almagest* eclipses, including another analysis of the eclipse of 19 March 721 B.C. Interspersed among these analyses are various calculations for dates in Mayer's own time and an investigation of Bullialdus's lunar tables. In the preface to his published lunar tables of 1753, Mayer highlighted the

---

[27] MS Mayer 15$_{41}$. ff. 140v–143v, 114r–146r, 155v–156v, 159r–166r.
[28] MS Mayer 15$_{41}$, f. 160r.

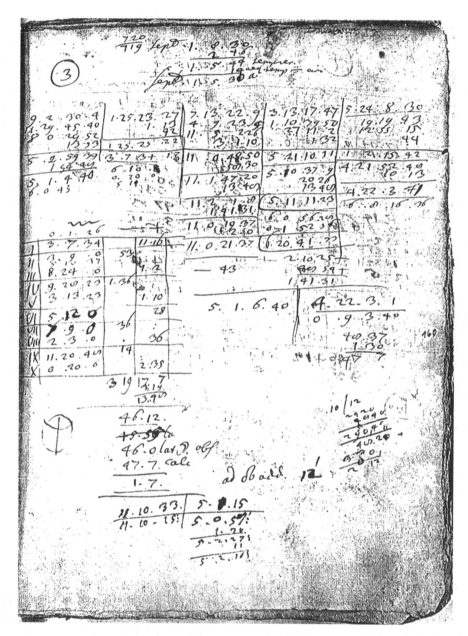

Fig. 7.2 Mayer's analysis of the eclipse of 1 September 720 B.C. (MS Mayer 15₄₁, f. 160r) (Courtesy Niedersächische Staats- und Universitäts-Bibliothek, Göttingen)

solar eclipses observed by Ibn Yūnus as being especially important for the determination of the secular acceleration, and even quoted the whole of Schickard's translation from Curtius's *Historia Coelestis*. Mayer also copied Schickard's translation into his manuscript notes.[29] Mayer analysed these eclipses using a refinement of the technique he had applied to the lunar eclipses reported by Ptolemy. He began by deducing the times of the observations from the altitude measurements

---

[29] MS Mayer 15₄₁, f. 140v.

given in the report. Mayer states in his manuscript notes that Cairo is at a latitude of 30°2′30″ and a longitude 1h56′ east of Paris. I do not know where he obtained these values from: they are close to but not exactly those determined by Chazelles (see Chap. 6). Mayer then calculated from his tables the longitude of the centre of the sun and moon at the observed time of the beginning of the eclipse. Next, he added the semi-diameters of the moon and the sun to the sun's longitude. If the sun and moon are at the same latitude, this addition should give the longitude of the moon since at the beginning of the eclipse the two luminaries are just touching. In theory a correction will need to be made to take into account the non-zero latitude of the moon during an eclipse. It is not clear from Mayer's notes whether he made this correction. This calculated lunar longitude and the longitude of the moon calculated directly from the tables could then be compared to give the error in the tables. It is not certain that Mayer calculated the error for these eclipses, however. He appears instead to have been content to accept the close agreement of the observed account with his calculations as evidence of the correctness of his tables, and therefore also as evidence for his estimate of the size of the secular acceleration. However, Mayer's analysis of the eclipse of 13 December 977 A.D. on MS Mayer 15_{41} f. 141r contains a mistake that to some extent weakens his conclusion. In calculating this eclipse, Mayer added a correction of 4′14″ for the secular equation to the mean longitude of the moon. This correction is considerably lower than that given by either his preliminary value for the coefficient of the secular equation of 6″ per century^2 or the 6.7″ per century^2 adopted in his 1753 tables. Using Mayer's preliminary coefficient of 6″ per century^2, however, gives a correction of 5′13″38‴, which rounds naturally to 5′14″. It seems likely, therefore, that Mayer made a simple mistake copying 5′14″ as 4′14″, a mistake that propagated throughout his analysis of this eclipse. The resulting error does not vitiate any conclusion Mayer drew from this eclipse, but it does weaken it.

Mayer's analysis of the lunar eclipses and occultations in the *Almagest* follows the pattern of the example described above with the single exception of the eclipse of 22 December 383 B.C. When he analysed that eclipse, Mayer, perhaps accidentally, omitted the correction for the moon's secular acceleration, and, although he derived an error in his tables of –40′ for this eclipse, he scribbled a note that this eclipse and particularly its duration need further examination. According to Ptolemy the eclipse began about half an hour before sunrise and Dunthorne had used this fact to place a crucial limit on the size of the secular acceleration. As noted by other astronomers both before and after Dunthorne, however, it is difficult to make this eclipse fit with the other eclipses in the *Almagest* no matter what corrections are made to the lunar theory. Probably for this reason, Mayer ignored this eclipse in his further analysis of the secular acceleration.

For each of the eclipses reported by Ptolemy that he analyses, Mayer starts with the local time of mid-eclipse at the place of observation. Mayer did not record the source for these times: some of the times are identical with the times Ptolemy derives for the moment of mid-eclipse (usually by converting from seasonal to equinoctial hours and estimating the duration of the eclipse from its magnitude), but in other cases they vary by as much as half an hour. Mayer probably went through the same process as Ptolemy of converting seasonal to equinoctial hours and estimating the duration, but I cannot explain the size of some of the differences between Ptolemy's and Mayer's derived times. An error of half an hour in the time of mid-eclipse would translate into an error of about 16′ in longitude, which is comparable in size to the errors he was finding between the observations and his tables. I can only conclude that Mayer was either extremely careless with his interpretation of Ptolemy's account or was deliberately misinterpreting some of them, though I can see no motivation for doing so.

In his analyses of these eclipses, Mayer also had to take into account the difference in longitude between where they were observed and Paris, the meridian of his tables. Mayer took the latitude of Babylon to be 32°30′ and its longitude 2h46′ east of Paris,[30] and Alexandria to be 1h52′ east of Paris (I could not find a note of its latitude in his manuscripts).[31] The longitude of Alexandria agrees with

---

[30] MS Mayer 15_{41}, f. 159r.
[31] MS Mayer 15_{41}, f. 164r.

Fig. 7.3 Mayer's spreadsheet analysis of observations in the *Almagest* (MS Mayer 15₄₁, f. 163v) (Courtesy Niedersächische Staats- und Universitäts-Bibliothek, Göttingen)

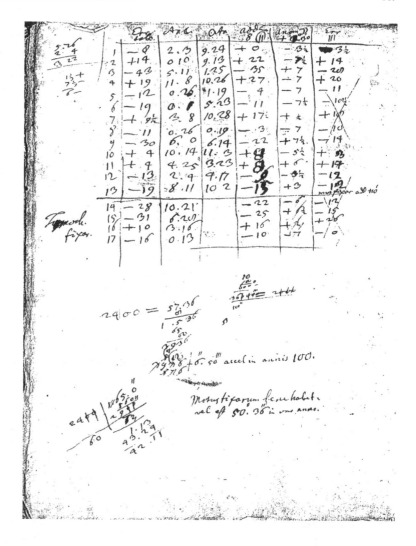

the value found by Chazelles and reported (with a typographical error) in the *Histoire de l'Academie Royale des Sciences Depius 1686 jusqu'a son Renouvellement en 1699*. Again, no evidence exists to indicate from where Mayer obtained the longitude of Babylon, but his value is close to the correct figure of about 2h48′ east of Paris.

On f. 163v Mayer drew a small spreadsheet in which he studied the errors found from 17 of the eclipses and occultations recorded in the *Almagest* (Fig. 7.3). The first column of the spreadsheet gives an index number for the observation which corresponds to number written within large circles at the top of each of his analyses of the *Almagest* observations (see Table 7.3). The first 13 observations are 9 of the 10 Babylonian eclipses (omitting the eclipse of 22 December 383 B.C.) followed by the 4 eclipses observed around Ptolemy's own time. Mayer's analysis of the early Greek eclipse observations in the *Almagest* appears in the following folios and so he must have studied them after he had compiled this spreadsheet. The final 4 lines of the spreadsheet, which are separated from the first 13 by a horizontal ruling, refer to the occultations observed by Timocharis and reported in *Almagest* VII 3. The second column of the spreadsheet contains the error in Mayer's tables for each observation. The errors agree with the values derived in Mayer's analyses of these observations contained in the earlier folios except for occasional rounding. Columns three and four give the anomaly of the moon and the sun respectively. The final three columns contain the corrected error if the

Table 7.3 Mayer's analysis of ancient lunar eclipses and observations on MS Mayer 15$_{41}$ f. 163v

Index number	Date of observation	Observer	Type of observation	Error (')	Corrected error (')
1	19 March 721 B.C.	Babylonian	Lunar eclipse	−8	+0
2	8 March 720 B.C.	Babylonian	Lunar eclipse	+14	+22
3	1 September 720 B.C.	Babylonian	Lunar eclipse	−43	−35
4	21 April 621 B.C.	Babylonian	Lunar eclipse	+19	+27
5	16 July 523 B.C.	Babylonian	Lunar eclipse	−12	−4
6	19 November 502 B.C.	Babylonian	Lunar eclipse	−19	−11
7	25 April 491 B.C.	Babylonian	Lunar eclipse	+9½	+17½
8	18 June 382 B.C.	Babylonian	Lunar eclipse	−11	−3
9	12 December 382 B.C.	Babylonian	Lunar eclipse	−30	−22
10	6 May 133 A.D.	Ptolemy	Lunar eclipse	+4	+8
11	5 March 136 A.D.	Ptolemy	Lunar eclipse	+4	+8
12	20 October 134 A.D.	Ptolemy	Lunar eclipse	−13	−9
13	5 April 125 A.D.	Theon	Lunar eclipse	−19	−15
14	9 March 294 B.C.	Timocharis	Occultation	−28	−22
15	9 November 283 B.C.	Timocharis	Occultation	−31	−25
16	Not specified	Timocharis	Occultation	+10	+16
17	Not specified	Timocharis	Occultation	−16	−10

changes noted at the top of each column are made to the lunar tables: adding 8′ to the mean position of the moon in column five (actually applied slightly differently—see below), adding 1°30′ to the moon's anomaly in column six and the combined error if both changes are made in column seven. Various values in the final column are crossed out, seemingly to indicate that these two changes to the lunar tables did not result in smaller errors overall.

Mayer's +8′ correction to the mean position that he had applied in column five of the spreadsheet had been derived from the error that he had found for the eclipse of 19 March 721 B.C. (number 1 in his list). He seems to have initially intended to apply it uniformly to all the observations, but after writing in the revised error for the 13 eclipses, he changed the correction for the final four eclipses (those observed by Ptolemy) to +4′ and wrote the resulting error over the first revised error. For the occultations observed by Timocharis, Mayer applied a correction of +6′. Thus his correction was +8′ for the Babylonian observations (721–382 B.C.), +6′ for Timocharis's observations (295–283 B.C.), and +4′ for the observations from Ptolemy's time (A.D. 125–136). The correction reduced, but did not eliminate, the error in the tables. Mayer evidently considered this to be a correction to the secular equation. He persuaded this further in a calculation jotted just below the spreadsheet. In his analysis of the 721 B.C. eclipse, Mayer had used his preliminary value for the secular equation of 6″ per century[2] to give a correction to the moon's mean position of 57′36″. Adding the 8′ correction to this gives 1°5′36″, or expressed in seconds 3936″. The year of the observation was 2421 years before the epoch of the tables, which Mayer rounded to 24 centuries. Mayer now divided 3936″ by 24^2, which gives 6″ 46′″ 15″″, which Mayer rounded to find 6″ 50′″ as a revised value for the secular equation. Mayer did not immediately adopt this revised value, however. In his analysis of the early Greek eclipse observations reported in the *Almagest* given in the immediately following folios, Mayer continued to use his preliminary value of 6″ per century[2] (or perhaps a slightly lower value in the case of two of them, although this may be just an artefact of rounding in his calculations).

At the same time as he used the observations reported in the *Almagest* to refine his value for the coefficient of the secular equation, Mayer tried out a completely different approach to the problem. On folios 157v and 158r, Mayer calculated the elongation of the moon from the sun at the beginning of the years 700 B.C. and 0 using the tables in Bullialdus's *Astronomia Philolaica* and his own tables. Mayer presumably worked with the elongation rather than the moon's position so as to remove any discrepancy caused by a difference in the assumed rate of precession or in the solar models. For 700 B.C., Mayer calculated that the elongation was $2^s 24°23'17''$ from Bullialdus's tables and $2^s 23°20'30''$ from his own tables if no correction for the secular acceleration is made. Applying his preliminary value for the coefficient of the secular equation of 6'' per century^2 simply to the calculated elongation (i.e. not taking it into account when calculating the lunar anomaly corrections), the elongation increases to $2^s 24°19'6''$. This value is closer to value from Bullialdus's tables, but still a little low. Mayer concluded that a slightly bigger value of the coefficient of the secular equation is needed of between 6½'' and 7''. Mayer followed the same steps for year 0, again finding that his preliminary value for the coefficient of the secular equation of 6'' per century^2 is again too low. In this case he did not write down a revised value, but his calculations indicate it should be about 8½''. This value may have seemed unacceptably high to Mayer when all the other evidence he had gathered pointed to between 6'' and 7''.

Whilst Mayer's attempt to refine the coefficient of the secular equation by comparing positions calculated in antiquity using his tables and Bullialdus's tables may be easily understood on a mathematical level, it is a little harder to understand it on a conceptual level. Why should a comparison of calculations made using a set of tables compiled in the seventeenth century with calculations made using Mayer's eighteenth-century tables provide a method for determining the error of Mayer's tables for dates in antiquity? I believe that the only explanation is that Mayer assumed that Bullialdus's tables were accurate for dates in antiquity (and by extension not accurate for dates at Bullialdus's own time) because they were derived from ancient observational data. In essence, Mayer treated Bullialdus's tables as if they *were* ancient tables, not seventeenth-century tables, because of their reliance of ancient observations, in which case his use of the tables to refine the secular equation was a justifiable approach.

Immediately following the analysis of Bullialdus's tables and the *Almagest* eclipse and occultation reports in MS Mayer 15_{41} are two unlabeled tables for the secular equation using a revised value for the coefficient of 6'' 40''' per century^2, which Mayer equated with 6.7'' per century^2 (Fig. 7.4). Folio 166v contains a table calculating the value of the secular equation at intervals of one century, together with various jotted calculations for constructing the table. Folio 167r contains a four-column table, again with a few jotted calculations in the margins. The first two columns give the year and value of the secular equation calculated based upon the table on folio 166v rounded to the nearest minute. In the third column, Mayer calculated the values of a linear function that intersects with the secular equation at zero in A.D. 1700 and 700 B.C. The fourth column contains the difference between the values in the second and third columns. It seems likely that Mayer was here investigating whether a simple linear function provided a sufficiently accurate approximation to the correct quadratic secular equation, and concluded that it did not. The values of the secular equation given in the second column are identical to those published in Mayer's 1753 tables, except for one obvious typographical error in the published version.

Mayer applied this new value for the secular equation to an analysis of the solar eclipse observed by Theon of Alexandria in the next folio (168r) of MS Mayer 15_{41}. Dunthorne had previously used this eclipse to place limits on the size of the secular acceleration and had (correctly) interpreted the time of the eclipse to be given in seasonal hours, even though the Basel edition of Theon specified that they were in equinoctial hours. Mayer initially followed the Basel edition and took the times to be in equinoctial hours when calculating the positions of the sun and moon at the time of the eclipse. It seems that Mayer subsequently realized the error in the Basel edition and at the bottom of his calculation added a note giving the time of the eclipse in equinoctial hours. He gives exactly the same derived time, 3h18', as Dunthorne had given in his 1749 paper, which raises the possibility that Mayer's correction was made after reading Dunthorne's paper. At the bottom of the page, Mayer scribbled the note "vid infra". A second attempt at calculating this eclipse is given in ff. 227v–228r. In this later analysis Mayer used this corrected time of the eclipse, but strangely seems to have forgotten to include the secular equation in his calculations.

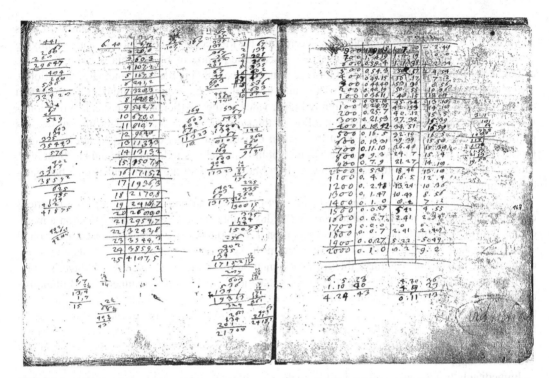

Fig. 7.4 Draft tables for the secular equation of the moon (MS Mayer 15_{41}, ff. 166v–167r) (Courtesy Niedersächsische Staats- und Universitäts-Bibliothek, Göttingen)

From this study of Mayer's manuscripts it is possible to summarize Mayer's work on the secular acceleration as follows. Mayer began by considering the mean motion of the moon in 60 years in his own time and from various other astronomical tables that were based upon ancient astronomical data to conclude that the moon was accelerating at a rate of about 4'''' per year, which corresponds to a coefficient of the secular equation of about 6" per century^2. Mayer then tested this value by calculating the circumstances of the two solar eclipses observed by Ibn Yūnus, finding sufficiently good agreement to convince himself of the accuracy of his tables (despite probably making a mistake in his calculation of the secular equation for one of these eclipses). He then undertook a similar comparison with the eclipse observations found in the *Almagest* and the occultations of Timocharis. At the same time, he compared his astronomical tables against those of Bullialdus for ancient dates. From these analyses of the *Almagest* eclipses, the occultations, and Bullialdus's tables, Mayer concluded that his preliminary value of 6" per century^2 was too small: he derived a value of 6" 50''' per century^2 from the *Almagest* data and a value of between 6½" and 7" per century^2 from Bullialdus's tables. Somehow from these he arrived at the figure of 6" 40''' per century^2, which he rounded to his final value for the coefficient of the secular equation of 6.7" per century^2. Finally, he subsequently tested this value against the eclipse of Theon and observations from the century before his own time.

Mayer's account of his determination of the size of the moon's secular acceleration to his account in the preface to his 1753 tables was ambiguous about the details of his method. From this analysis of his manuscripts we can see that Mayer had good reason not to give a full account of his work. Rather than having a clear procedure for deriving the size of the acceleration from the observational data, Mayer's method relied on an intuitive refinement of a preliminary estimate through comparison with historical data. This refinement was not achieved through a rigorous, step-by-step process, but rather by combining evidence from different approaches to different types of data, both historical observations and earlier lunar theories, in a way that Mayer knew led to a better final value, but could never formally justify, or even easily explain. Mayer faced essentially the same problem with the refinement of many of the parameters of his lunar tables, which he had derived by a process of iterative correction through the examination of the errors in the calculated positions of large bodies of observed lunar positions. Mayer did not attempt to explain how he had derived the coefficients of the equations of anomaly in the preface to the 1753 tables, for which he was strongly criticised by d'Alembert. In his *Theoria Lunæ*, Mayer justified the need to derive some parameters from observational data rather than theory, but again skirted the issue of how he had done so. As Wepster has recently shown, Mayer *did* develop a method for determining these parameters using his spreadsheets, but the method relied on Mayer's skill and judgement in handling large datasets and could not be explained to or repeated by others.[32] Mayer's determination of the size of the secular acceleration and the coefficient of the secular equation relied on similar techniques, but was even less formalized as he had to combine evidence from different types of data in order to refine his preliminary value.

The account given in the preface to the 1753 tables, however, is not simply ambiguous, it is misleading, although carefully worded so as to never be factually incorrect. But the clear implication in the preface is that Mayer distrusted Ptolemy as a source of ancient data, that he instead relied upon the two solar eclipses observed by Ibn Yūnus to determine the size of the secular acceleration, and that finally he confirmed his result by comparing values for the mean motion of the moon in 60 years found in different sets of lunar tables. But as I have described, the process was almost exactly the opposite: Mayer started by looking at the values for the mean motion of the moon in different lunar tables to provide a preliminary estimate for the size of the secular acceleration and then refined this estimate through his analysis of the eclipses from Ptolemy and more work on earlier lunar tables. The solar eclipses of Ibn Yūnus, which Mayer highlights in the preface to the 1753 tables, were apparently only used to confirm that that there was a secular acceleration and that the preliminary value found from the analysis of earlier lunar tables was plausible, and were not used to refine that value. It seems likely that Mayer presented his work in this way because he wished to downplay the importance of the eclipses from Ptolemy in his determination of the size of the secular acceleration. Mayer rightly commented that the *Almagest* eclipses were crudely observed and did not all agree well with his tables, and drew attention to the charge that Ptolemy had altered the times of some of the eclipses. As I will discuss in the next section, Mayer needed to criticize Ptolemy in order to argue against Euler's claim of a solar secular acceleration. I suspect that Mayer's account of the derivation of the moon's secular acceleration was influenced by this need to present Ptolemy as an unreliable source.

Although Mayer's tables were appreciated for their accuracy, he was criticized by some astronomers for not establishing their theoretical basis. He attempted to rectify this by writing a theoretical explanation of the lunar theory, the *Theoria Lunæ juxta Systema Newtonianum*, which was completed and sent to the Board of Longitude in London by the end of 1755, but only eventually published in 1767. In it Mayer argued that some parameters of the lunar theory could not be derived theoretically and could only be found empirically. The size of the moon's secular acceleration seems to have been one of these parameters and Mayer says nothing about it in the *Theoria Lunæ juxta Systema Newtonianum*. In a letter to Euler dated 25 November 1753, Mayer considered the possibility that the moon's secular acceleration might be caused by the resistance of the æther to the moon's motion, as Euler had suggested, but also wondered if the cause was not instead something to do with the sun's gravity:

> acceleration in the Moon's motion may now also be best accounted for by this resistance. Yet I think that the cause might also lie in the attraction of the Sun, from which the rest of the inequalities in the Moon's motion originate. For just that reason, some of these inequalities become appreciable, because their periods are very long; thus it could happen that in the combination of so many angles, if the approximation would be carried very far, an inequality would be found whose period may be very long and [even] infinitely large. And in these

---

[32] Wepster (2010), pp. 143–176.

instances, this inequality would transform itself into a continuous acceleration. Meanwhile, it appears to be very difficult or quite impossible to derive this [conclusion] from the theory itself.[33]

As soon as the 1753 tables were published, Mayer returned to refining the parameters of the equations of anomaly and began drafting new sets of lunar tables, several of which are preserved in manuscript form. Of those that I have consulted, none contains a table for the secular equation. This is almost certainly simply because Mayer considered the secular equation in the 1753 tables to be sufficiently accurate; some of the preserved manuscript tables contain only those tables which had been revised (generally the anomaly tables), and omit those tables which were the same as in an earlier set (for example, the mean motion tables). However, Mayer's "final manuscript tables", published by Maskelyne with revisions by Bradley after Mayer's death, does contain a table for the secular equation (Fig. 7.5). Interestingly, this table is based upon a coefficient for the secular equation of $9''$ per century^2, not the $6.7''$ per century^2 of his 1753 tables. Two copies of this revised table for the secular equation in Mayer's hand are preserved among the papers of Nevil Maskelyne in the Royal Greenwich Observatory archives: an untidy, scribbled draft and a neat copy.[34] These sheets were sent to England along with Mayer's so-called final manuscript tables by Mayer's widow after his death and were written at an unknown date after 1755. I could find no other trace of this higher value for the coefficient of the secular equation among Mayer's other preserved manuscripts, and so it is not possible to determine exactly how he arrived at this higher figure. It is highly likely, however, that he did not obtain this value through a re-evaluation of the ancient data. In the final lunar tables Mayer adopted a higher value for the mean motion of the moon than in his 1753 tables: $1^s10°44'9''$ in 60 years instead of $1^s10°43'30''$ in the 1753 tables. A higher mean motion at the 1700 epoch would require a greater secular acceleration in order to still be compatible with the ancient data. Mayer almost certainly deduced the higher value from the secular acceleration from a consideration of his revision of his value for the moon's mean motion, not from a reanalysis of data on which he had derived his earlier value for the secular acceleration.

Mayer on the Secular Acceleration of the Sun

Mayer began his work on the lunar theory using Euler's 1746 lunar tables. As I have discussed in Chap. 4, Euler's tables included a secular equation for the calculation of the sun's position. Euler argued that as the Earth moves on its orbit it will be subject to a very small resistance caused by the æther. This resistance would cause the radius of the Earth's orbit to gradually decrease, shortening the period of the orbit which would be seen from the Earth as an acceleration in the sun's motion. In the first of two letters to Caspar Wetstein that were published in the *Philosophical Transactions* in 1752, Euler claimed that his conclusion was supported by Bernard Walther's solar observations from the fifteenth century. In his second letter, Euler reaffirmed his belief in an acceleration in the sun's motion, but lamented that the ancient observations were not reliable enough to prove it:

But in order to put this Fact out of Doubt, we ought to be furnished with good ancient Observations, and also to be very sure of the Time elapsed, since those Observations, to this Day: Which we are not, with regard to the Observations that PTOLEMY has left us. For Chronologists, in fixing the Moments of those Observations, run into a Mistake, by supposing the Sun's mean Motion to be known; which ought rather itself to be determined by those same Observations. Now if we reduce the Days marked by PTOLEMY to the Julian Kalendar, we run the Risque of committing an Error of a Day or two, in the whole Number of Days elapsed, from that to our Time; because the Course of the Julian Years, according to which every fourth ought to have been Bissextile, has been

---

[33] Mayer to Euler 25 November 1753; translation by Forbes (1971), p. 77.

[34] MS RGO 4.125, p. 29 and an unpaginated sheet.

[L]

Acceleratio motus medii L U N. Æ,

Addenda femper ad longitudinem Lunæ mediam.

Anni æræ vulgar.		+ o ' "		Anni æræ vulgar.	
Poft C. N.	1700	0 . 0 . 0,0		1700	Poft C. N.
	1720	0 . 0 . 0,4		1680	
	1740	0 . 0 . 1,4		1660	
	1760	0 . 0 . 3,2		1640	
	1780	0 . 0 . 5,8		1620	
	1800	0 . 0 . 9,0		1600	
	1820	0 . 0 . 13,0		1580	
	1840	0 . 0 . 17,6		1560	
	1860	0 . 0 . 23,0		1540	
	1880	0 . 0 . 29,2		1520	
	1900	0 . 0 . 36,0		1500	
	2000	0 . 1 . 21		1400	
	2100	0 . 2 . 24		1300	
	2200	0 . 3 . 45		1200	
	2300	0 . 5 . 24		1100	
	2400	0 . 7 . 21		1000	
	2500	0 . 9 . 36		900	
	2600	0 . 12 . 9		800	
	2700	0 . 15 . 0		700	
	2800	0 . 18 . 9		600	
	2900	0 . 21 . 36		500	
	3000	0 . 25 . 21		400	
	3100	0 . 29 . 24		300	
	3200	0 . 33 . 45		200	
	3300	0 . 38 . 24		100	
	3400	0 . 43 . 21		0	
	3500	0 . 48 . 36		100	Ante C. N.
	3600	0 . 54 . 9		200	
	3700	1 . 0 . 0		300	
	3800	1 . 6 . 9		400	
	3900	1 . 12 . 36		500	
	4000	1 . 19 . 21		600	
	4100	1 . 26 . 24		700	
	4200	1 . 33 . 45		800	
	4300	1 . 41 . 24		900	

Fig. 7.5 Mayer's table for the secular equation in his final lunar tables (Mayer, *Tabulæ Motuum Solis et Lunæ Novæ et Correctæ*, table L) (Courtesy John Hay Library, Brown University Library)

frequently interrupted by the Pontifices; of which we find some sure Marks in CENSORINUS and DION CASSIUS. Wherefore it might well happen, since the Times mark'd by PTOLEMY, that there has really been a Day or two more than we reckon, and consequently, that PTOLEMY'S Equinoxes, out to be put a Day or two back; which would lengthen the Years of those Times.[35]

Mayer, however, rejected Euler's argument for a solar secular acceleration. In the preface to his 1753 tables, Mayer explained that there was no evidence for a change in the length of the year:

As far as the solar tables are concerned, which are offered here together (with the lunar tables), although in them others have left less for me to do, there was no single element that I have not again confirmed by observations. Thus the motion of the sun's apogee such as offered by Flamsteed's tables has been retained since it is in keeping with the observations of Hipparchus and Albategnius. Moreover, the quantity of the tropical solar year is set as a constant, on the guidance of the observations of Hipparchus, Albategnius, Walther, and, in one word, all in whom faith and authority can be placed, with the single exception of Ptolemy. Without any doubt, his equinoxes, however, were not deduced from heaven, but rather were fitted to his tables and to the quantity of the solar year which he had received from Hipparchus; as I will clearly explain elsewhere. I am quite convinced by some very powerful arguments that the chronological order of time from Ptolemy to today has not been disturbed by any loss of a day or two.[36]

By examining the historical observations of the times of equinoxes, Mayer found that the length of the year had remained constant since antiquity. Only the equinoxes which Ptolemy claimed to have observed disagreed with this finding. Like Longomontanus and others before him, Mayer believed that Ptolemy had forged these observations in order to confirm Hipparchus's value for the length of the year. Euler had suggested that the discrepancy between Ptolemy's equinoxes and those from Hipparchus and other ancient astronomers might have been caused by errors of a day in the conversion of ancient dates to the Julian calendar, but Mayer knew that this argument did not make any sense because all the dates of lunar observations recorded in Ptolemy's *Almagest* were clearly correct when converted into the Julian calendar. An error of a day in the calendar would correspond to an error in lunar longitude of about 13°, which could not be explained by any correction to the lunar or solar theory. Therefore any error in the calendar around the date of an equinox would have had to have happened shortly before the equinox and then be corrected shortly afterwards, which Mayer dismissed as an unreasonable assumption.

Mayer developed his arguments against a solar secular acceleration further in a letter he wrote to Euler on 22 August 1753:

Your view that the solar years may be unequal has always appeared so fundamental to me that I would not have departed from it if I had not been compelled to do so through a more accurate investigation both of the ancient observations and of the correctness of the common chronological time-scale. It is first of all quite certain that from the time in which the observations of Hipparchus and Ptolemy were made, neither more nor fewer days have passed than one is normally used to count.[37]

Mayer explained that he could not make his solar and lunar tables compatible with both the equinoxes and the eclipses which Ptolemy claimed to have observed, and if any changes to the tables were made in that direction then it would throw out the agreement with the medieval Arabic observations and the modern observations, concluding that "it was not possible for me to even bring the observations from the two most recent centuries into agreement with it, without committing an error of half or even a whole hour",[38] when these eclipses were accurately observed to within a few minutes. "If you want to take these arguments, which I mention only briefly, into more careful consideration", he continued,

---

[35] Euler, "Concerning the Contraction of the Orbits of the Planets", pp. 356–357.

[36] Mayer, "Novae Tabulae Motuum Solis et Lunae", pp. 391–392.

[37] Mayer to Euler 22 August 1753; translation by Forbes (1971), p. 73.

[38] Mayer to Euler 22 August 1753; translation by Forbes (1971), p. 74.

"I do not doubt that you would not completely vindicate me".[39] The error, Mayer said, must lie instead with the Ptolemaic equinoxes:

> Since, therefore, the time-reckoning requires no intercalary [day], one must necessarily conclude that the Ptolemaic equinox is really in error. My opinion of it is this. Ptolemy has simply based his solar tables on the equations and observations of Hipparchus. This is revealed, among other [reasons], since his previously-mentioned observations of the equinox are on the whole newer than his observations of planets. He had also brought the planetary tables into order, before he had observed his equinoxes. That could not be done, however, without first of all having the solar tables in a correct [form]. He consequently borrowed the solar motion from Hipparchus without any particular investigation. This, however, put the length of the solar year at $365^d\ 5^h\ 55^m$ …, so large in fact, *quod probe notandum*, that it is [deduced] from its luni-solar cycle of 76 years, and not from actual observations.[40]

Mayer explained that Hipparchus' year-length is too large by about 6½ minutes, which accumulates to about 1¼ day in the 300 years between Ptolemy and Hipparchus, and this 1¼ day is the error by which Ptolemy's equinoxes are too late. Thus, Ptolemy simply calculated the times of his equinoxes by adding Hipparchus's year-length to the Hipparchian equinoxes. In Mayer's view, Ptolemy was aware of the error in the times of these equinoxes, but did nothing about it because he would then have had to go back and correct all of his planetary theories:

> It can be that Ptolemy perceived this error of his solar tables in his observations of the equinoxes, which are the very last of all his remaining observations; only, because he had already built his whole system upon it, perhaps he had rather wanted to discard his observations than to attempt to revise his system from the outset. Since, however, no one could object to it, he pretended that the erroneous equinoxes of his tables were true and observed. There are more and newer examples of an astronomer, from too great a love for his constructions, falsifying observations. This is certain at least in Lansberg and Riccioli. How much more simply could not Ptolemy, who perhaps did not imagine that one would ever be able to disclose his deception through more accurate observations, have fallen into error.[41]

Mayer continued by explaining that the conclusion that Ptolemy's equinox data is derived from his solar model is supported by Ptolemy's value for precession. Ptolemy "pretends that the motion of the fixed stars is approximately 1¼° too small, namely just as much as his tables err in the motion of the Sun".[42] Mayer argued that Ptolemy observed the positions of the fixed stars by measuring the distance of the star from the moon and the distance of the moon from the sun. Ptolemy then calculated the sun's position using his solar model ("since at that time he had still not observed his equinox"),[43] and consequently the longitude of the sun, the moon, and the fixed star were all found to be about 1¼° too small. Thus, Mayer argued, the error in Ptolemy's value for precession came from the error in his solar model. He concluded, "This circumstance is sufficient to show that no credit should be attached to Ptolemy's observations of the equinoxes".[44]

Euler replied to Mayer on 25 February 1754, after he had seen Mayer's tables in the *Commentarii Societatis Regiae Scientiarum Gottingensis*, saying that while he still believed in the resistance of the æther to the motion of heavenly bodies, he accepted Mayer's arguments against any evidence for there being a secular acceleration of the sun:

> That you have also by means of the Sun found my lunar equation to be valid has also pleased me not a little, and regarding the resistance of the aether (which still appears to me well established and necessary) I would be rather dubious since I must confess that the discovered accelerated motion of the mean Moon could easily be

---

[39] Mayer to Euler 22 August 1753; translation by Forbes (1971), p. 74.

[40] Mayer to Euler 22 August 1753; translation by Forbes (1971), pp. 74–75.

[41] Mayer to Euler 22 August 1753; translation by Forbes (1971), p. 75.

[42] Mayer to Euler 22 August 1753; translation by Forbes (1971), p. 75.

[43] Mayer to Euler 22 August 1753; translation by Forbes (1971), p. 75.

[44] Mayer to Euler 22 August 1753; translation by Forbes (1971), p. 76.

the effect of an inequality whose argument may have a period of many centuries. With regard to the Sun, I am completely convinced of my error; it is nevertheless to be regretted that one has no older and more accurate observations.[45]

Mayer never published his promised demonstration of Ptolemy's faking of his equinox observations, although he incorporated many of the arguments in his letter to Euler into § 30 of his unpublished treatise "Vorlesung über Sternkunde",[46] albeit with somewhat more measured language (he dropped the references to Lansberg and Riccioli faking observations, for example). This claim, however, was central to his rejection of Euler's proposed solar secular acceleration. Mayer was convinced that there was no evidence for an acceleration of the sun's (really the Earth's) motion, but in order to prove it he needed to show that the ancient value for the length of the year found by Hipparchus and confirmed by Ptolemy did not reflect the true length of the year at that time, and this could only be done by showing that Ptolemy's equinox observations were faked. This placed Mayer in a difficult position because, as I have discussed, Ptolemy's lunar observations played an important role in Mayer's determination of the size of the lunar secular acceleration. Thus, Mayer had to assume that Ptolemy's solar observations were in error but that the lunar observations were more reliable. It is probably for this reason that Mayer highlighted the role of the solar eclipses seen by Ibn Yūnus in his determination of the secular acceleration, and implied that he had not used the eclipses reported by Ptolemy, when the opposite was closer to the truth. The only other option for Mayer would have been to try to explain why Ptolemy's lunar observations were reliable when his solar ones were not—an argument it would not have been easy to justify.

Mayer's Use of Historical Evidence

Mayer's use of historical eclipse observations to investigate the moon's secular acceleration may be contrasted with Dunthorne's approach described in the preceding chapter. Whereas Dunthorne had avoided the problem of the accuracy of ancient timing methods and the ambiguities in Ptolemy's descriptions of eclipses by developing a technique for deriving limits to the size of the secular acceleration from the visibility of selected eclipses near the horizon, Mayer first treated the ancient observations as if they were accurate and reliable observations from which the time of syzygy could be determined, and then looked for reasons to reject results which seemed too discordant to the other observations. Mayer was apparently fairly widely read in the history of astronomy: he cited Bullialdus's remarks on Ptolemy's eclipse records in the preface to his 1753 tables, and references to Bullialdus, Riccioli, Curtius, and Longomontanus appear in his manuscript notes. Mayer's reading of Bullialdus and Longomontanus no doubt influenced his view that Ptolemy was an unreliable source.

Shortly after the publication of his 1753 tables, Mayer received from Euler copies of four letters sent by the Jesuit Antoine Gaubil in Beijing to the astronomer Joseph Nicholas Delisle.[47] These letters, three of which are dated 22 October 1752 and the fourth 2 November 1752, concern two main topics: observations made by Gaubil in Beijing to be used to determine the longitude difference between Beijing and Paris, and the dates of solar eclipses described in the "*Tchou chou*" (*Zhushu jinian* 竹書紀年) and the "*Chou king*" (*Shujing* 書經). The eclipse in the *Shujing*, said to have taken place in the ninth month of the fifth year of a king Zhong Kang, had been proclaimed as the most ancient astronomical observation known throughout the world by the Jesuits who had communicated

---

[45] Euler to Mayer 26 February 1754; translation by Forbes (1971), p. 79.

[46] MS Mayer 9; edited by Forbes (1972).

[47] Mayer's copies of these letters are preserved in MS Mayer 15_2, ff. 1r–3v.

descriptions of its sources to a European audience. The most commonly accepted date for the eclipse was 12 October 2155 B.C., although William Whiston proposed 22 October 2137 B.C. and other astronomers had suggested other dates. The *Zhushu jinian* seemed to contain a report of an eclipse on 28 October 1948 B.C., but the date preserved in the text was clearly corrupt, and Gaubil concluded that three 60-year cycles needed to be added to the date to give 13 October 2128 B.C., suggesting a significant revision of *Zhushu jinian* chronology.

Gaubil wrote that he had investigated whether eclipses had taken place on the various possible dates with the help of "a Mandarin Chinese highly skilled in calculation, highly recommended by P. Gogails".[48] Gaubil's Chinese assistant had concluded that a solar eclipse would have been seen in Beijing between 6h1' and 7h34' after midnight on the 12 October 2155 B.C., and so at the place of observation "from the calculations of the Mandarin Chinese (a good Christian), it follows that it was clearly visible".[49] This seemed to confirm the identification of the eclipse in the *Shujing*. For the *Zhushu jinian* eclipse, Gaubil said that he had rejected the eclipse of 13 October 2128 B.C. as a possible date for this eclipse because using the lunar tables available to him he had found that this eclipse would not have been visible in the appropriate part of China. However, he asked Delisle to investigate more carefully whether there could have been an eclipse on this date, as he felt that the *Zhushu jinian* was in other aspects a reliable text.

Euler had hoped that the ancient Chinese observations might have been useful for refining the lunar and solar theories, but expressed his frustration over how Gaubil had reported them:

> I am taking the liberty of communicating to you a letter from China which was itself sent to me from England, from which it is to be seen how the Jesuits have wanted to correct the ancient epochs through their lunar tables. Notwithstanding, it is now to be wished that they had reported the old observations properly without their corrections, so that you could perhaps still have drawn useful information from them.[50]

Euler's criticism of Gaubil was misplaced as the account of the eclipse Gaubil gave was fairly faithful to the original text.

Mayer's response to receiving the copies of Gaubil's letters was to calculate the circumstances of the eclipse of 13 October 2128 B.C. using his own tables. Mayer's calculations are preserved in MS Mayer 15_{41}, f. 371v (Fig. 7.6). He began by investigating the position of the sun and moon for 13 October 2128 B.C. at 0 hour (noon epoch). Mayer took the time difference between China and Paris as 7 hours and calculated the mean longitudes of the sun and moon for −2127 October 12 at 17 hours obtaining $6^{s}3°7'40''$ for the sun and $6^{s}5°43'9''$ for the moon. He then added a correction for the secular equation of the moon of $2°43'35''$ to obtain a final mean longitude for the moon of $6^{s}8°26'17''$. This value for the secular equation is calculated using Mayer's final value for the coefficient of $6.7''$ per century[2] as given in Mayer's 1753 tables. Mayer then changed the time of his calculation to 3h30' earlier, and adjusted the moon's mean longitude by the corresponding amount ($-1°55'17''$), to obtain a final mean longitude for the moon of $6^{s}6°31'27''$. Mayer made a similar adjustment to the sun's position, but mistakenly read the value for 4h30' rather than 3h30' from his table, and obtained $6^{s}2°56'35''$ for the sun's mean longitude. Finally, with these mean longitudes, positions of the apogee and the none, Mayer calculated the true positions by applying the various equations of anomaly to arrive at final positions for the sun and moon of $6^{s}2°1'30''$ and $6^{s}2°11'23''$ respectively, a difference of only $9'53''$, implying that conjunction took place close to 3h30' before noon (or 8h30' after midnight) on 13 October 2128. At the bottom of the page Mayer calculated more exactly that conjunction took place at 8h42'0'' after midnight, but also that the moon's latitude at that time was 58' north of the ecliptic. On the facing page, Mayer scribbled other possible dates for the eclipse that were in the

---

[48] MS Mayer 15_2, f. 1v.

[49] MS Mayer 15_2, f. 2r.

[50] Euler to Mayer 26 February 1754; translation by Forbes (1971), pp. 79–80.

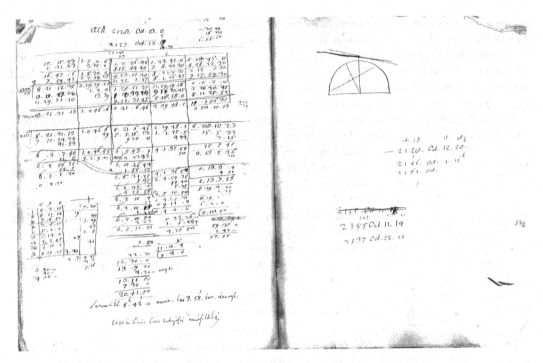

Fig. 7.6 Mayer's analysis of the visibility of the solar eclipse of 13 October 2128 B.C. in China (MS Mayer 15$_{41}$, ff. 371v–372r) (Courtesy Niedersächische Staats- und Universitäts-Bibliothek, Göttingen)

same Saros series as the eclipse of 11 October 2155 B.C. or 13 October 2128 B.C., estimating their times by assuming that the Saros was about 8 hours longer than a whole number of days.

On 6 March 1754 Mayer replied to Euler thanking him for sending Gaubil's letters and explaining that according to his tables the eclipse of 13 October 2128 B.C. was invisible in China:

> I return my most sincere thanks for the communication of the letter of P. Gaubil; as soon as I have studied it with proper attention, I shall send it back to you. Meanwhile, I can state so much, that according to my tables the solar eclipse of the year 2128 B.C. was invisible in the whole of China. The Chinese reports seem to me very suspect, or at least to require a severe criticism.[51]

Mayer never returned the letters to Euler and no reference to them appears in any of their subsequent correspondence, so far as it is preserved. Mayer did, however, discuss the eclipse again in his unpublished "Vorlesung über Sternkunde".[52] In § 54 Mayer related how he had received Gaubil's letters from Euler and recounted their contents. He described Gaubil's correction of the *Zhushu jinian* chronology and explained the he had recalculated whether an eclipse could have been seen in China on 13 October 2128 B.C.:

> I have taken the effort to calculate this eclipse according to my tables, though I was almost assured well in advance that my efforts would be of no particular utility. According to this calculation I find for two reasons that this eclipse was invisible throughout China. In the first place the latitude of the moon was overly to the north for the shadow or the penumbra to have touched even the smallest part of China; and secondly the time of conjunction of the moon with the sun fell just before the rising of the sun. I have full reason to believe that this calculation is more correct than those which indicate that the eclipse was visible in China, because in my tables, the motion of the moon was slower in former times than now, which others do not have. And so I have grounds

---

[51] Mayer to Euler 6 March 1754; translation by Forbes (1971), pp. 84–85.

[52] MS Mayer 9; edited by Forbes (1972).

that the entire report of the seen and yet nonetheless invisible solar eclipse is a highly suspect and completely false account. Perhaps a Chinese astronomer in a much later time has brought out a solar eclipse from a very imperfect backwards calculation of the abovementioned year, and with it given that it appeared real at that time. Ignoring that, it appears as well from the above that even this year of the eclipse is still highly doubtful. [53]

It is interesting that Mayer draws attention to the fact that his tables include a correction to the moon's position to take into account the moon's secular acceleration. However, Mayer's claim that the eclipse would have been invisible in China because conjunction took place before sunrise seems at odds with the calculation that it would be at 8h42′ after midnight found in the manuscript discussed above (sunrise in October would have been only a short while after 6 am). In the next section, Mayer had the following to say about the eclipse on 12 October 2155, which had been widely identified with the report in the *Shujing* and had been suggested also as the *Zhushu jinian* eclipse:

> The same P. Gaubil, concerning this eclipse, when it is set in the year 2128 before the birth of Christ, himself found some problems, and one could just as well specify the eclipse which occurred in the year 2155 B.C. as the one reported in the Tschu-schu (*Zhushu*). The common tables indeed give for 12 October of this year a morning eclipse, and make it visible at Peking (Beijing) and other localities of China. But with my tables the time of the true new moon falls in the 4th hour after midnight, and so this eclipse would have been completely invisible. I conclude from all this that a sharp judgement is necessary of the report that is brought to us from China. A half-scholarly and in astronomical matters moderately experienced people like the Chinese were before the Europeans came to them, could easily forge such false eclipses, and to make their origins respectable and their erudition old supply inaccurate calculations for the observations; because they believed that he was not smart enough to discover the fraud.[54]

Mayer may have been over-zealous in his criticism of Gaubil and the Chinese, but he showed good judgement in rejecting the various dates proposed for these eclipses. It must have been tempting to Mayer to attempt to use the eclipse to confirm the accuracy of his tables and to refine his value for the secular acceleration. The secular equation increases with the square of the number of centuries since the epoch of the tables, and so the great antiquity of this eclipse would allow the coefficient of the secular equation to be determined with high precision. However, in order to use the eclipse in this way, its date needed to be known with certainty. As since the date was determined by looking for the best fitting eclipse calculated using existing theory, the process would have been circular.[55] To his credit, Mayer resisted this temptation, and rightly placed his faith in his tables over over-interpretations of this ancient record.

Other parts of Mayer's unpublished "Vorlesung über Sternkunde" provide further insight into his knowledge of the history of astronomy and his view of ancient astronomical sources. Mayer began writing this treatise on 9 December 1753, returning to it over the next year until completing the final section on 17 December 1754.[56] The treatise is not written to be a complete history of astronomy but rather a historical exploration of topics that were of concern to Mayer at the time. Prominent among these topics are ancient knowledge of the length of the year and the synodic month and the recording of eclipse observations in antiquity, which were clearly related to Mayer's investigation into secular accelerations of the sun and moon. Mayer divided the "Vorlesung" into five parts: (1) the origin of astronomy, (2) the first attempts to correct the length of the year and the months, (3) year-lengths in antiquity, (4) the invention and division of the zodiac, and (5) eclipses. The planets and the fixed stars received no attention whatsoever. In writing the treatise, Mayer used material from a variety of classical sources, including Herodotus, Pliny, Diodorus Siculus, and Josephus, all of which he probably

---

[53] MS Mayer 9, § 54; edited by Forbes (1972), p. 97.

[54] MS Mayer 9, § 55; edited by Forbes (1972), p. 97.

[55] Robert Newton has called this approach "playing the identification game" and demonstrated how it results in unjustifiable confirmation of existing parameters. See Newton (1969).

[56] Forbes (1972), pp. 17–18.

read first-hand,[57] but also referred at one place to Curtius' *Historia Coelestis* and Weidler's *Historia Astronomiae*, both of which he may well have drawn on for further information and ideas.

Most of Mayer's "Vorlesung" is fairly standard material for histories of the time, with one interesting slant: Mayer places a large emphasis on ancient Egyptian astronomy and shows almost no interest in ancient Babylon. The reason seems to be Mayer's interest in ancient knowledge of the length of the year, and he saw the Egyptian 365-day calendar as an important stage in the development of an accurate measurement of the tropical year. Classical sources provided little information on Babylonian knowledge of the length of the year. But even when discussing the Saros cycle of 223 months, commonly called the "Chaldean period", Mayer preferred to see an Egyptian rather than a Babylonian origin:

> It is not very certain what people have made this very important discovery. Some write that it is the Chaldeans, but others, and perhaps with more reason, the Egyptians; as from out of their hieroglyphs, sacred and profane customs and also other details, we have traces of enough that they must have been far ahead in astronomy than other nations of their time.[58]

That there was no evidence for the Egyptians having known of the Saros did not worry Mayer because

> If we know so little with certainty about their discoveries, it appears such is not because of their ignorance, but rather to be attributed to the habit that the Egyptian priests, or call as you like their scholars, to make their discoveries secret, and those who were not of their order, carried them forward under dark mysteries and hidden images.[59]

Greek astronomers received a fairer treatment than their Babylonian counterparts. Mayer discussed Thales as the first person to predict a solar eclipse in advance (although he stressed that Thales had learnt his astronomy in Egypt), the tradition of Greek calendrical astronomy including Meton and Callipus, and Hipparchus's and Ptolemy's solstice and equinox observations and their determinations of the length of the year. As I had discussed above, Mayer was highly critical of Ptolemy's solar theory and his faking of his equinox observations. At the very end of the treatise, Mayer described the eclipse records in Ptolemy's *Almagest*, but did not make any critical comments on them.

Apart from Mayer's peculiar obsession with an imaginary Egyptian astronomy, Mayer's treatment of ancient astronomy in the "Vorlesung" is in keeping with his use of ancient astronomical observations in his studies of the secular accelerations. The focus on year and month lengths ties in with Mayer's methods for determining whether there were secular accelerations from changes in the lengths of these fundamental intervals between antiquity and Mayer's own time. Mayer's criticism of Ptolemy and of the Chinese eclipse accounts agreed with his stated position elsewhere, even though, as I have discussed above, Mayer relied upon Ptolemy's eclipse reports much more heavily that he was willing to admit.

In conclusion I would summarize Mayer's approach to historical observations as intuitive and essentially straightforward. Mayer may have used the historical data in imaginative ways, such as his comparison of values for the mean motion of the moon found in different sets of tables to investigate the moon's secular acceleration, but he analysed the historical observations using exactly the same method as he analysed contemporary observations. He made no attempt to extract additional data from the observations than that which is clearly stated in the historical source. But Mayer's intuition and good judgement came into play when he compared the different observations and chose how to combine the various data sets—and which ones to ignore—to obtain a final answer.

---

[57] Forbes (1972), p. 17.

[58] MS Mayer 9, § 51; edited by Forbes (1972), p. 95.

[59] MS Mayer 9, § 51; edited by Forbes (1972), p. 95.

Chapter 8
The Final Synthesis: Jérôme Lalande

Lalande ... demonstrated an early love for fame which was at all times his ruling passion, and which he sought to satisfy by any means which were presented or could be imagined.

—Jean Baptiste Delambre, *Histoire de l'Astronomie au Dix-Huitième Siècle* (Paris 1827), p. 547.

Dunthorne and Mayer had both firmly established the existence of the moon's secular acceleration, but differed in their determination of its size. Furthermore, neither Dunthore nor Mayer had provided a fully justified term for the effect of the moon's secular equation in their lunar tables: Dunthorne gave a detailed description of how he had derived the coefficient of the secular equation, but presented it as a correction to his existing lunar tables. Mayer, by contrast, had fully incorporated the moon's secular acceleration into his lunar tables, but had not given a detailed explanation of how he had derived its size. Further confusion existed over whether the sun's motion was also subject to a secular correction. Euler had reasoned that resistance to the motion of a heavenly body by the æther would cause such an acceleration in the Earth's motion and had included a secular term in his solar tables, but Mayer claimed that the ancient observations provided no evidence for a solar equation. Such was the state of the subject when astronomy's great opportunist, Jérôme Lalande, made a bid to have his name attached to the definitive solution to the problem. It was evident that determining whether the sun and moon were subject to a secular acceleration, and if so, what was its magnitude was going to become an important problem both for the construction of accurate astronomical tables—something that was gaining in importance because of the role of lunar distance measurements in determining geographical longitude—and in testing the ability of Newton's gravitational theory to explain the motion of bodies in the solar system. Lalande's paper on the secular equations, published in 1757, was successful in bringing the study of the size of the moon's secular acceleration using historical records to an end for the next several decades. As I will show, however, Lalande was extremely lucky in this outcome as the arguments presented in his paper scarcely justify the faith placed in his result.

J.M. Steele, *Ancient Astronomical Observations and the Study of the Moon's Motion (1691–1757)*, Sources and Studies in the History of Mathematics and Physical Sciences, DOI 10.1007/978-1-4614-2149-8_8, © Springer Science+Business Media, LLC 2012

Jérôme Lalande

Joseph-Jérôme Lalande was born on 11 July 1732 at Bourg-en-Bresse in France.[1] He was educated first at the Jesuit College in Lyon and moved to Paris to study law. In Paris Lalande lodged at the Hotel de Cluny. Joseph-Nicolas Delisle, professor of astronomy at the Collège Royale, had an observatory in the same building, and after meeting Delisle, Lalande was able to sit in on Delisle's astronomy lectures and Lemonnier's physics lectures at the Collège. Through these contacts, Lalande was sent to Berlin to participate in the measurement of the parallax of the moon and of Mars. Lalande was nothing if not an opportunist, and in Berlin he made contact with Euler, Maupertius, and other leading scholars and was admitted as a member of the Academy, a huge honour for the young Lalande.[2]

Lalande returned from Berlin and was immediately welcomed into French scientific society (he was elected to the Académie des Sciences in 1753). Lalande spent the remainder of his life in Paris, except for two lengthy trips to England and one to Rome. He was appointed editor of the *Connaissances des temps* in 1760, and in 1762 he succeeded Delisle as professor of astronomy at the Collège Royale. Lalande possessed a highly accomplished political sense and worked to make himself one of the most influential scientists in France during the second half of the eighteenth century, carefully manoeuvring through the Revolution and surviving the rise of Napoleon unscathed and still a powerful figure.

Lalande's reputation was built on his political acumen, his shameless quest for publicity, and his skill as a remarkably clear author who could communicate astronomical ideas to a wide audience. His *Astronomie* and *Astronomie des Dames* were influential and widely read books that ensured Lalande's name was not forgotten. Late in his life, Lalande was appointed director of the Paris Observatory, where he completed an enormous catalogue of stars which he published as *Histoire Céleste Française* in 1801. Lalande died on 4 April 1807.

Lalande's Mémoire sur les Équations Séculaires

On 19 November 1575 Lalande presented his "Mémoire sur les Équations Séculaires, Et sur les moyens mouvemens du Soleil, de la Lune, de Saturne, de Jupiter, & de Mars, Avec les observations de Tycho-brahé, faites sur Mars en 1593, tirées des manuscripts de cet Auteur" to the Académie Royale des Sciences for publication in their *Mémoires*. A mixture of review and Lalande's own research (not always clearly distinguished from one another), the paper was intended to settle the question of whether the motions of the sun, moon, and the outer planets were subject to any secular accelerations, and if so, to determine the size of the coefficient of the secular equation in the calculation of that body's longitude. Related to these questions was the determination of the mean motion of a heavenly body at a given epoch, since if the body is subject to an acceleration (or a deceleration) then the mean motion of that body will change over time. Lalande began his paper by explaining that although this problem has been known for quite some time, no consensus has yet been reached:

> The period of the celestial revolutions & the size of the mean motions of the planets are not the same as now than formerly; it is true that it has been discussed for over a century, but it was only little discussed, and over it astronomers were not in accord.[3]

---

[1] For detailed biographical information, see Amiable (1889) and Dumont (2007).

[2] Gapaillard (2010).

[3] Lalande, "Mémoire sur les Équations Séculaires", p. 411.

This provided the justification for Lalande to make his own contribution.

Lalande's paper is divided into a substantial introduction in which he reviewed the history of the discovery of the secular accelerations and their investigation by past and contemporary astronomy followed by five "articles" dealing in turn with the secular acceleration of the sun, the moon, Jupiter, Saturn, and Mars. A 25 page appendix (almost half the paper) contains an edition of Tycho Brahe's observations of Mars made at Hven during the year 1593 based upon a manuscript in the library of the Académie Royale des Sciences. Lalande used these observations in his article on Mars.

In the introduction Lalande gave a fairly extensive overview of the history of work on the secular acceleration. He began with Kepler's claim of changes in the mean motion of the outer planets, Flamsteed's confirmation that Saturn was decelerating and Jupiter accelerating, Halley's discovery of the moon's secular acceleration and the secular equations in his tables for Jupiter and Saturn, and Euler's claim that there is a solar secular acceleration. This last acceleration is, of course, really an acceleration in the motion of the Earth around the sun, which has important consequences:

> This acceleration of the Earth has already given rise to dire consequences for humanity as we are almost announced the time and the way it should end: indeed, if the Earth accelerates its motion, it is a certain proof that it feels a resistance from the æther ... This case having begun to act, it always would act, the distance that the Earth to the Sun would not cease to fall, because the speed would always be tested against a new resistance, the effect would become more and more considerable, as the Earth approached the centre, because the density of light & the central force both increase as the square of the distance decreases: thus by degrees the Earth descends to the Sun, for to be absorbed and destroyed ... We will depart from such melancholy forebodings, even if we be deprived of the resulting proof in the creation and against the eternity of the world; these purely human proofs are useless to the Christian philosophers.[4]

Having spent more than a page explaining how the existence of a secular acceleration of the sun implies a grizzly end of the world, the Earth falling into the sun's fire, Lalande only now came to his punchline: "We shall see that the acceleration of the Earth does not exist".[5]

Lalande then turned to the effect on the motion of a heavenly body caused by a constant acceleration. He demonstrated using simple mathematics that the effect of the acceleration on a body's longitude increases with the square of time and commented on the problem of using the period of daily rotation of the Earth as a unit of time which, he says, we have no way to determine whether it has been uniform over the centuries.

Lalande on Ptolemy and the Secular Acceleration of the Sun

Lalande began his study of the secular equations with sun. As I have discussed in Chap. 4, Euler proposed on theoretical grounds that the sun's motion was subject to a secular acceleration and he therefore included a secular equation in his solar tables. Mayer, however, claimed that the preserved ancient observations did not provide any evidence for this solar secular acceleration and constructed his solar tables with a constant mean motion. Establishing whether or not the sun's motion is accelerated and what is the correct value for the sun's mean motion is essential for the construction of any astronomical tables because the sun's motion is needed to calculate the anomalies in the motion of the moon and the planets. Consequently, the sun was the natural place for Lalande to begin his study of the secular equations. Lalande's study of the solar secular equation in the first article of his paper is of particular interest as it provides important insights into Lalande's methods of working with historical data and, in particular, his opinion of Ptolemy.

---

[4] Lalande, "Mémoire sur les Équations Séculaires", pp. 413–414.
[5] Lalande, "Mémoire sur les Équations Séculaires", p. 414.

Table 8.1 Cassini's analysis of historical equinox data

Observer	Year length
Hipparchus	365d5h48′49″
al-Battānī	365d5h48′49″
Regiomontanus and Walther	365d5h48′51″
Copernicus	365d5h48′43″
Landgrave of Hesse and his Mathematicians	365d5h48′49″
Tycho	365d5h48′47″
Riccioi, at Bologne	365d5h48′35″
Cassini, at Bologne	365d5h48′53″
Observations made at the Paris Observatory	365d5h48′46″

In order to determine whether the sun was subject to a secular acceleration, Lalande considered the length of the tropical year derived from observations made at different epochs. Lalande's discussion of the historical data was based upon Jacques Cassini's analysis in his *Éléments d'Astronomie* of 1740:

It can be seen in the comparison made by M. Cassini of the ancient observations with the modern of the equinoxes of Hipparchus, Albategnius, the prince of Cassel & Tycho Brahe, that they give the same length of the year, within 2″ of time; this already proves the uniformity of the motion of the Earth.[6]

Cassini had compared all of the historical equinox observations known to him with equinoxes observed in Paris between 1672 and 1739 to determine the mean length of the year between the ancient observations and Cassini's own time.[7] For each group of observations Cassini determined the average value for the year length and summarized the data on p. 232 of the *Éléments d'Astronomie* (see Table 8.1). He omitted the year length obtained from Ptolemy's observations since "they depart by one minute from all the others".[8] Cassini concluded that the length of the tropical year was 365d5h48′47″. Lalande argued that because the year lengths derived from the observations of Hipparchus, al-Battānī, the Landgrave of Hesse, and Tycho Brahe were all within 2′ of Cassini's value, the length of the year had not changed over time, which in turn meant that the sun was not subject to a secular acceleration.[9] Cassini's comparisons of the other observations gave almost the same result, except for the equinoxes observed by Ptolemy.

But what of Ptolemy's observations? Lalande noted the suggestion (by Euler) which appeared in the *Philosophical Transactions* of an error in the calendar and summarized the evidence from Dion Cassius and Censorinus which told of mistakes in the assignment of leap years within the Julian calendar. He explained that a 1-day error in the calendar would show up as a 13° error in lunar position, which cannot be reconciled with the ancient lunar observations. Instead, Lalande attributed the problem to Ptolemy himself:

The most natural way to solve the difficulty is to totally reject Ptolemy's observations; everything seems to indicate that this author, attached to assumptions that could not yet be then very accurate, attempted to make all the rest consistent with them; not only were his own observations adjusted to the theory, but he even admit he has made some changes in the time of eclipse observations, as noted by M. Boulliau. Between the eclipse which arrived in the year 547 of Nabonassar in the month *messori*, & that of the year 548 in the month *mechir*, he assumes 178 days 6h50′ mean time, while Hipparchus reckoned 50′ less. Between this latter eclipse & the eclipse of the month *messori* 548, he has 176 days & 24′, whereas Hipparchus reckoned 1h20′, that is to say, that Ptolemy has boldly added 56′ to the interval of these observations.[10]

[6] Lalande, "Mémoire sur les Équations Séculaires", p. 418.
[7] Cassini, *Éléments d'Astronomie*, pp. 207–232.
[8] Cassini, *Éléments d'Astronomie*, p. 232.
[9] See also Wilson (1985), pp. 61–62 and Chapin (1988), pp. 185–186.
[10] Lalande, "Mémoire sur les Équations Séculaires", p. 420.

Lalande explains that this conclusion is also confirmed by Ptolemy's equinox observations themselves, which imply an inconsistent year length, and thirdly by Ptolemy's incorrect value for the rate of precession. It is worth noting that Lalande's first (including the reference to this particular chapter in Bullialdus) and third reasons are also given by Mayer in the preface to his 1753 tables. Lalande refers to Mayer in the second article of his paper, but not here, and one must wonder whether Mayer's explanation was on his mind when he wrote this passage.

Lalande is not finished with Ptolemy, however. Having concluded that Ptolemy's equinox observations should be rejected, and suggested that Ptolemy fiddled with some of the times of earlier eclipses he reports, Lalande then reports on his own study of Ptolemy's value for the lunar parallax, published in the *Mémoires* for 1752, which showed that Ptolemy's value was too great, and on Kepler's study of Ptolemy's value of the obliquity of the ecliptic. "Thus", he wrote, "almost all astronomers have found Ptolemy flawed, each in the part he has (examined) in depth; is this not sufficient grounds for dismissing the observations of this author, when we find ourselves with the impossibility of reconciling them with the ancient ones which he reports?".[11]

Lalande on the Secular Acceleration of the Moon

Article 2 of Lalande's "Mémoire sur les Équations Séculaires" concerns the mean motion and the secular acceleration of the moon. Lalande began by outlining the history of the discovery of the secular acceleration by Halley and the subsequent investigations by Dunthorne and Mayer. Lalande's account, which ignores Whiston's discussion of the size of the acceleration and Struyck's rejection of the existence of the acceleration, became the standard presentation of this history by almost all later authors.[12] He began:

> We have already seen that M. Halley was the first who had suspected a physical acceleration in the motion of the moon, as can be seen in his notes on the observations of Albategnius, written in 1693, & on the relationship between the ruins of Palmyre in 1695. But to make this research useful, it was necessary to have the longitudes of the places where Albategnius observed; I think that after the observations that M. de Chazelles made in Alexandrette in 1694, M. Halley concluded more positively this acceleration, at least if we are to believe a passage in the second edition of M. Newton.[13]

As far as I am aware, Lalande is the first and only author to have suggested that Halley used de Chazelles's longitude determinations to reinvestigate the moon's acceleration. The passage in Newton (quoted in Chap. 2) certainly provides no indication of this. But it would not be surprising if Halley did return to the topic as improved determinations of the longitude of cities in the near east became available.

Lalande next speculated on the reasons for the omission of the passage referring to the moon's acceleration in the third edition of Newton's *Principia* (see Chap. 2), before turning to Dunthorne and Mayer:

> In any event, M. Halley does not explain anything further on this subject, which is what provided a place for M. Dunthorne, & later M. Mayer, to examine the same matter, the one & the other concluded that there is an acceleration; but also there is a difference in their results, and M. Mayer has hidden all his calculations, & M. Dunthorne made use of tables which he had not constructed, but only shown the elements of in another volume of the Philosophical Transactions, so the matter seems to me capable still of some discussion.[14]

---

[11] Lalande, "Mémoire sur les Équations Séculaires", p. 421.

[12] For example, the accounts in Delambre (1827), pp. 597–598 and Grant (1852), p. 60 are clearly based upon Lalande. It is quite likely that Lalande was unaware of the contributions of Whiston and Struyck.

[13] Lalande, "Mémoire sur les Équations Séculaires", p. 426.

[14] Lalande, "Mémoire sur les Équations Séculaires", pp. 426–427.

Lalande was certainly correct in criticising Mayer for not giving details of how he determined the size of the moon's secular accelerations. It is also true that Dunthorne's analysis was based upon tables that were not fully published. But these tables included only a small correction, described in his 1746 *Philosophical Transactions* paper, to his published tables of 1739, and Dunthorne clearly expected his readers to be able to make the necessary adjustments. As we will see, Lalande's criticisms are rather disingenuous as he himself was guilty of both hiding the details of his analysis and of using corrected, and therefore unpublished, versions of tables.

Lalande began his discussion with the two solar eclipses observed in Cairo by Ibn Yūnus:

> The observations which are most decisive in this matter are two eclipses of the sun observed near Cairo in the year 977 & 978 by Ibn Junis. This astronomer laboured at celestial observations by order of the Caliph Abu-Haly-Almanzor the Wise, who commanded Egypt. Skikardus said that the tables of this author were in the hands of Golius, Professor at Leiden, & they contain many observations, from his time and earlier times. M. de l'Isle has obtained a copy of the same Arabic manuscript from M. Luloss, correspondent of the Academy at Leiden, & he makes us hope for a translation. These two eclipses are reported in the Prolégomenes de l'Histoire Céleste de Tycho, and they are almost the only ancient observations which we are know the time accurately, because they observed the altitude of the sun at the beginning & the end of each eclipse.[15]

Interest among astronomers in the observations contained in Golius's manuscript of Ibn Yūnus had been continual since the announcement of the manuscript's existence. It would not be surprising if Delisle had attempted to have the work translated, although I know of no evidence for the copy obtained by Luloss reaching Paris.

Lalande next summarized the two observations and deduced the local times of the beginning and end of each eclipse from the measurements of the sun's altitude:

> Year 367 of the Hegire, Thursday 28, *Rabie II*, that is to say the fourth month of the year, & this year of the Saracens began on 19 August 977, the beginning of the eclipse happened when the altitude of the sun was 15°43′, it ended when the sun had an elevation of 33½°, the eclipse was of 8 digits, this corresponds to 13 December 977, beginning at 8h24′24″, end at 10h43′4″, assuming the latitude of Cairo is 30°2′30″. Saturday 29 of the month *Sywal* or *Sylwal* (that is the tenth month or the Paschal month) there was again an eclipse of 7 digit & a half, the sun at the beginning had an elevation of about 56°, at the end of 26°, this corresponds to 28 June 978, beginning 2h31′, end 4h50′, true time.[16]

The Julian date of the second eclipse contains a typographical error: 28 June should read 8 June (the correct date is given in the next paragraph). In Table 8.2 I compare Lalande's derived local times of the beginning and end of each eclipse with the times derived by Mayer and Dunthorne. There is close but not perfect agreement between the three sets of times. Mayer and Lalande both take the latitude of Cairo to be 30°2′30″; Dunthorne does not specify the latitude, but it is likely to be close to this as well. For the first eclipse, the times given by Lalande and Mayer are identical except for a discrepancy of 40 seconds in the time of the end of the eclipse (Lalande 4″ versus Mayer 44″). It is tempting to see Lalande's 4″ as typographical error for 44″. Dunthorne's values for the beginning and end agree with Mayer's when rounded to the nearest minute. For the second eclipse, Lalande and Dunthorne gave the same times in minutes; Mayer's are more precise, quoted to the second, but this precision is spurious when we consider that the altitude measurements are in whole numbers of degrees. It is possible that Lalande simply copied the times from Mayer and Lalande, but since Mayer and Lalande, and probably Dunthorne, were working with the same latitude for Cairo, it is to be expected that they would independently derive the same, or at least very close, local times from the altitude measurements.

Lalande continued: "From this it can be concluded that on the 12 December 977, 19h21′, mean time at Paris, the moon had a longitude of 8ˢ26°19′, and on the 8 June 1h24½′, mean time at Paris, 2ˢ22°16¾′".[17] The times presumably refer to the moment of conjunction, but I cannot follow exactly

---

[15] Lalande, "Mémoire sur les Équations Séculaires", p. 427.

[16] Lalande, "Mémoire sur les Équations Séculaires", p. 427.

[17] Lalande, "Mémoire sur les Équations Séculaires", p. 427.

Table 8.2 A comparison of the local times of the eclipses observed by Ibn Yūnus deduced by Lalande, Mayer and Dunthorne

Date	Contact	Solar altitude	Local time		
			Lalande	Mayer	Dunthorne
13 December 977	Start	15°43′	8h24′24″	8h24′24″	8h25′
	End	33½°	10h43′4″	10h43′44″	10h45′
8 June 978	Start	56°	2h31′	2h30′16″	2h31′
	End	26°	4h50′	4h50′24″	4h50′

how Lalande obtained them. If we simply assume that the time of conjunction is midway between the beginning and the end of the eclipse, apply a correction for the difference in longitude between Cairo and Paris of 1h58′20″ taken from Chazelles's survey, and correct for the equation of time, we obtain noon-epoch mean times at Paris of about 19h31′ and 1h41′ respectively (Lalande had not previously specified that the local times of the eclipse of 12 December 977 were in the morning and the local times of the eclipse of 8 June 978 were in the afternoon, but this is obvious from the increase in solar altitude during the course of the first eclipse and decrease in solar altitude during the second eclipse). However, because of parallax the midpoint of a solar eclipse does not necessarily coincide with conjunction and it seems likely that Lalande made a correction for this effect when deriving his mean time of conjunction. The lunar positions at these times were almost certainly determined by calculating the solar position for these moments using a version of Cassini's solar tables which Lalande had adjusted for the slightly longer length of the tropical year that he had derived in the preceding article. Next, Lalande says that he has calculated the moon's position for these times using Clairaut's lunar tables, finding positions which were too high by 21 1/3′ for the first eclipse and 15¾′ for the second. Because the first eclipse took place near perigee and the second almost at apogee, Lalande says that a close agreement between the two errors would indicate that the position of the apogee given in the tables is correct. He writes:

> a difference of 4 minutes in the place of the apogee produces a difference of 1 minute between the observations, so that the two eclipses that I have reported, determine the place of the apogee with greater precision than it is possible to hope; if we increase by 4 minutes the place of the apogee, than the tables of M. Cassini give for the tenth century, we will make the two errors perfectly equal.[18]

He continues:

> Assuming that the epoch is well defined in the tables of M. Clairaut, this reduces the secular motion of the apogee by 48 seconds, an extremely slight difference in such research. So the secular movement of the apogee in one hundred Julian years will be 3ˢ19°13′28″, greater by 2′13″ from what M. Mayer has in his tables. As for the error of the Tables, having first reduced the mean motion, & assuming this for the sun as I determined in the first part of this memoir, we have the longitude being too large by 17′50″, the motion in the tables being too great for these two important observations at a rate of 2′18″ per century.[19]

Lalande here proposes altering the motion of the moon's apogee in Clairaut's tables to 3ˢ19°13′28″ per century (confusingly for our purposes he calls this the "secular movement", but he is using the term "secular" simply to mean the motion over time, not an acceleration in the motion), and noted that this is bigger than the corresponding value in Mayer's tables. Lalande presumably adjusts the calculation of the moon's position to take into account this new value of the motion of the apogee, and concludes that the calculated longitudes (now of both eclipses since he has adjusted the motion of the apogee to make the errors equal) are 17′50″ too large. This would imply that the moon's mean motion per century was too great by 17′50″ divided by the number of centuries between Ibn Yūnus

---

[18] Lalande, "Mémoire sur les Équations Séculaires", p. 428.

[19] Lalande, "Mémoire sur les Équations Séculaires", p. 428.

and the epoch of the tables (say 1750). The result would be $2'17''42'''$, which Lalande rounds to $2'18''$. Lalande ignores the effect of parallax in computing these errors. As Dunthorne explained in detail, because of parallax, there is not a linear relationship between the time of a solar eclipse and the solar and lunar longitude. In other words, correcting the moon's position by the error he has just deduced will not give precisely the longitude of the sun and moon at conjunction. The error is small, but not negligible.

In this analysis of Ibn Yūnus's eclipses, Lalande has assumed that the moon's motion is not accelerated, but that instead the mean motion given in Clairaut's tables is slightly too high. In order to confirm this result, however, he needs more data:

> but unfortunately, this point is the only one we have with any certainty in all antiquity, the observations which have passed through the hands of Ptolemy being suspect, as we have from later remarks of his own admissions, & agreeing also badly between them: however, as there is a fairly large difference between the mean motion that results from the old observations and the ones we have just determined, it does not appear that is can be attributed to the corrections that Ptolemy might have made to the observations from his hypotheses. Let us choose the most ancient eclipse, which appeared to him the most respectable, as it was already very ancient for him.[20]

The mean motion for the moon derived from the eclipse observations described by Ptolemy, however, does not agree with the mean motion Lalande has just derived from the observations of Ibn Yūnus. Lalande had already demonstrated Ptolemy's untrustworthiness—and was at pains to highlight it again—but even allowing for Ptolemy having fudged the observations to make them agree with his models, he found it impossible to make the two mean motions agree. This would imply that the moon's motion had changed over time—in other words, that it was subject to an acceleration. Lalande says he will demonstrate this by looking at the oldest eclipse in the *Almagest* as he believes that it is the report that Ptolemy is least likely to have altered out of respect for its great age, a very peculiar and unjustifiable reason. If Ptolemy is as wont to tamper with observational reports as Lalande claims, one wonders why he would have had a special respect for this particular observation and not, for example, for the eclipse observed 6 months later?

Lalande's analysis of this eclipse is interesting and worth quoting in full:

> 19 March 721 B.C. under the fourth of the Chaldean kings who reigned after Nabonassar, the moon began to be eclipsed the whole of one hour after it rose. To deduce the true time, it is necessary to know the situation of Babylon, which for fifteen hundred years no longer exists, and some place on the Tigris, others on the Euphrates, and whose remains have not been recognized by travellers. On the map of Persia by M. Delisle, it is found at 32 2/3° of latitude, which differs very little from Baghdad, which is on the Tigris: but a passage from Ptolemy persuades me that it was more to the north than M. Delisle placed it.
>
> Ptolemy, reporting an observation made near the winter solstice 313 B.C. shows the time of sunset at 4h48'; I remove 2' for the effect of refraction; and suppose the obliquity of the ecliptic to be 23½°, it follows that the latitude of Babylon was 36°10'; to find only 33½° it would be necessary to assume that Ptolemy has the length of the day too great by a quarter of an hour, and that the obliquity of the ecliptic would have been 24°; assumptions which would be forced.
>
> Assuming, then, the declination the moon 4½° north, the rising of the moon 5h39' at Babylon, the semi-duration of the eclipse 1h54', the difference of meridians 2h32', the equation of time 10 min, additive to the true time, we have the conjunction 6h11', mean time at Paris. If we employ with the tables of M. Clairaut the mean motion found from the two preceding observations, we find the longitude too small by 1°27'; which is a clear indication of an acceleration.[21]

Establishing the location of the ancient city of Babylon was essential in order to make use of the Babylonian observations. Dunthorne had accepted Ptolemy's statement that Babylon lay 50' east of Alexandria, which when combined with Chazalles's value for the longitude of Alexandria placed Babylon 2h41'13" east of Paris, and used Ptolemy's latitude for the city. Mayer had assumed that the

---

[20] Lalande, "Mémoire sur les Équations Séculaires", pp. 428–429.

[21] Lalande, "Mémoire sur les Équations Séculaires", p. 429.

latitude of Babylon was 32°30′ and its longitude 2h46′ east of Paris. Lalande found that on Delisle's map of Persia, Babylon was placed at latitude 32 2/3°, close to the true value. But Lalande was misled into using a higher latitude by his interpretation of a passage in Ptolemy concerning an observation made around winter solstice in the year "313 B.C.". The date is clearly a mistake for 383 B.C. and must refer to the observation of a lunar eclipse reported on the night 22/23 December 383 B.C. in *Almagest* IV.11. Ptolemy says here that "1 hour of night in Babylon is 18 time-degrees (for the night is 14 2/5 equinoctial hours long)".[22] This implies that sunset is at 4h48′ pm. Lalande reduces this by 2′ to take account of atmospheric refraction near the horizon and then uses this time to calculate that the latitude of Babylon is 36°10′, which places Babylon far too much to the north. It seems as if Lalande was attempting to demonstrate his cleverness by deriving Babylon's latitude in this way, but he was badly mislead by his acceptance of Ptolemy's value for the length of night at Babylon. This value is derived from the very simple ratio 3:2 for the length of day and night at the solstice and it is surprising, given his hostility to Ptolemy elsewhere, that Lalande accepted it without question. What is even more interesting, however, is that Ptolemy's value for the length of night is reported by Dunthorne in his discussion of this eclipse, and Dunthorne's paper contains the same typographical error in the year. It seems very likely, therefore, that Lalande simply took the information about this eclipse from Dunthorne's paper, and did not spot the misprint in the date. With this in mind, it is worth remarking that the eclipse of 19 March 721 B.C. was also discussed by Dunthorne. This makes Lalande's peculiar remarks that he trusted this eclipse because it was the oldest one reported by Ptolemy look very suspicious. Indeed, I suspect the true reason Lalande chose this eclipse is that it was quoted and analysed by Dunthorne, and so Lalande did not have to bother reading Ptolemy to analyse it.

Having deduced the latitude of Babylon, Lalande then returns to the eclipse of 19 March 721 B.C. Assuming that the eclipse began one hour after moonrise, Lalande first calculates that moonrise at Babylon on that day took place at 5h39′, added the hour to obtain 6h39′ for the beginning of the eclipse, then added 1h54′ for the semi-duration of the eclipse to give the moment of opposition. Ptolemy had assumed that the semi-duration of the eclipse was 2 hours, so this last value must be Lalande's own calculation. Finally, Lalande subtracted 2h32′ for the difference in longitude between Babylon and Paris (a very low value—the true difference is about 2h48′), and added 10′ for the equation of time to obtain the time of observed opposition as 6h11′ mean time at Paris. Using Clairaut's tables adjusted for the mean motion that Lalande had found from the Ibn Yūnus eclipses, he calculated that the moon's longitude by the tables was too small by 1°27′. This is too great a difference to be an observational error, so Lalande concludes that the moon must be subject to an acceleration.

Having shown that the moon's motion must be accelerated, Lalande now turned to the size of the acceleration. After a brief remark about the different interpretations of the 721 B.C. eclipse by Cassini and Dunthorne, which seems out of place in his discussion, Lalande gave his conclusion:

> The mean secular motion of the moon between the years 977 & 1700, must be increased by 1′11″47‴, for the mean motion in a century to be $10^s7°53′21″$; so we must necessarily introduce a secular equation into the tables of the moon.
>
> M. Mayer gives this equation 7″ per century, or 1°4′
> M. Dunthorne .... 10″ per century, or 1°36′ for the year 700 B.C.
> After the preceding calculation, we have 9.886″, or 1°35′

Lalande does not explain how he has arrived at these figures and his terminology is far from clear. 1′11″47‴ divided by the number of centuries between 977 and 1700 (7.23) gives 9″56‴ or 9.886″, which is the coefficient of the secular equation. But where does the value 1′11″47‴ come from? Using Lalande's secular equation, the secular correction to the moon's position in A.D. 977 is 8′38″

---

[22] Ptolemy, *Almgest* IV.11; translation by Toomer (1984), p. 212.

and for 721 B.C. is $1°37'$. Now Lalande had found that if he used the mean motion from the time of Ibn Yūnus to calculate the eclipse of 721 B.C., there was an error of $1°27'$. Adding the secular correction $8'38''$ to the $1°27'$ error for the 721 B.C. gives 1;35,38, close to the value for the 721 B.C. eclipse given by Lalande's secular equation. Thus it is clear that Lalande has derived the coefficient of his secular equation by combining the Ibn Yūnus and the 721 B.C. eclipses, although exactly how he did this is not clear.

In addition to stating his own determination of the coefficient of the secular equation, Lalande also gives the values deduced by Mayer and Dunthorne. He rounds Mayer's value from $6.7''$ per century^2 to $7''$ per century^2, but correctly gives the size of the secular correction for 700 B.C. from Mayer's table. Dunthorne's value of $10''$ per century is correctly reported from Dunthorne's paper, but Lalande has calculated the size of the secular equation for 700 B.C. himself assuming a constant acceleration from the A.D. 1700 epoch. However, Dunthorne, because he worked with a mean motion of the moon which was too low, had two zero points for his correction: one in A.D. 700 and the other in A.D. 1700. Because of this, Dunthorne's own value for the secular equation in 700 B.C. is only $56'$, somewhat lower than that given by Lalande. In effect Lalande has corrected Dunthorne's secular equation so that it fits the correct value of the mean motion at the A.D. 1700 epoch.

Lalande ended his discussion of the lunar secular acceleration by discussing the eclipse of Theon:

> We have also an eclipse of the sun observed by Theon on 6 June 364; the beginning at 3h18°, the end at 5h15°: the tables give just 11 minutes too much, which would make the secular acceleration much smaller; but this observation does not outbalance the two Arab observations we have just used, which seem much more exact, according to which the acceleration of the moon can be set at $10''$ per century.[23]

Lalande does not say where he has obtained the details of this eclipse from, but it would seem that once more he has taken them straight from Dunthorne. As I discussed in Chap. 6, Dunthorne had noticed that Theon's timings of the eclipse were given in seasonal not equinoctial hours, despite the Basel edition of Theon saying the opposite, and had converted the seasonal hours into equinoctial times. The times of the beginning and end of this eclipse given by Lalande match exactly the equinoctial times given by Dunthorne. Lalande notes that the eclipse of Theon seems to imply a much smaller secular acceleration than he has deduced from the other eclipses, but decides that Theon's observation is not as accurate as those made by Ibn Yūnus, and so he is happy to ignore this eclipse. Lalande concludes by saying that the "acceleration of the moon", which is actually the coefficient of the secular equation, is equal to $10''$ per century^2, which is exactly the value Dunthorne had found.

Despite his claim to want to place the study of the moon's secular acceleration on a firm footing, Lalande's study suffers from both the flaws he attributed to Mayer and Dunthorne: he did not provide a full explanation of either his methodology or his calculations, and he worked with unpublished modifications of existing sets of astronomical tables. Furthermore, it is clear that Lalande based his analysis on Dunthorne's paper of 1749. Lalande used a subset of the observations cited by Dunthorne and no others, adopted many of Dunthorne's interpretations, and ultimately derived the same value for the size of the coefficient of the secular equation as Dunthorne. Indeed, it is tempting to see Lalande's first value of this coefficient, $9.886''$ per century^2, as a phantom value constructed to be slightly different to Dunthorne's $10''$ per century^2, the figure that Lalande finally rounds his value to. At best, Lalande's work can be seen as a review and re-analysis of Dunthorne's work, influenced also by Mayer's stated high regard for the Ibn Yūnus observations, rather than as an independent investigation of the problem.

---

[23] Lalande, "Mémoire sur les Équations Séculaires", p. 430.

Lalande returned to the question of the moon's secular acceleration in his *Astronomie*.[24] In his discussion of the moon, Lalande briefly addressed the acceleration of its mean motion in four paragraphs numbered 1162–1165 in the first edition. In the first paragraph he very briefly explained what the secular equation of the moon was and why it was needed. The second paragraph gives a condensed version of the history of the acceleration's discovery and subsequent investigation, based upon the account in his 1757 paper. He immodestly concludes: "finally I discussed this matter with more care and details in the Memoirs of the Academy for 1757, *pag.* 426. Here in a few words is the outcome of my research about this",[25] before giving a very brief account of his 1757 study in the following paragraph. He says that the two solar eclipses observed by Ibn Yūnus are the most decisive for studying this matter and gives his result for the coefficient of the secular equation, $9.886''$ per century[2]. Interestingly he does not mention his use of the 721 B.C. eclipse from Ptolemy's *Almagest* in determining this value. In the final paragraph, Lalande makes an interesting and new comment: he casts doubt upon whether the two eclipses reported by Ibn Yūnus were observed or whether he had calculated them:

> I must however warn that M. Grischow when in Leiden in 1749, hired M. Schultens Professor of Arabic Language to research and translate the Arabic MSS from which these observations are taken; I saw in London in the hands of M. Dr Bevis this translation; it contains obscurities, and M. Bevis even thinks that they were rather calculations than actual observations; … people educated in Oriental Languages will not have a better opportunity to make their studies useful.[26]

Lalande wrote of his meeting with Bevis in the entry for Monday 18 April in his diary of his trip to England in 1763: "Lunch with Doctor Bevis, who gave me a more correct copy of the observations of ibn Junis".[27] Schulten's translation was sent to the Royal Society in London where it was later studied by Costard and others. Lalande had stressed the importance of these eclipses for his determination of the secular equation and so was obviously worried about whether the eclipses might have been calculated. Concern over whether this eclipse was observed or calculated was misplaced; they were observations and Lalande was correct to use these eclipses.

At the end of the first volume of the *Astronomie*, Lalande reproduced a set of astronomical tables. He chose Lacaille's solar tables and Mayer's 1753 tables for the moon, including Mayer's table for the secular equation using Mayer's coefficient of $6.7''$ per century[2]. Lalande had criticised Mayer's tables in his 1757 paper, claiming that the value for the motion of the apogee in a century in Mayer's tables was too small by $2'13''$ and that Mayer's coefficient for the secular correction was too small by almost $3''$ per century[2]. The fact remained, however, that Mayer's tables were more accurate than any other set of lunar tables available at the time, and had been widely recognized as such by the time Lalande was writing the *Astronomie*. Lalande had no choice but to include Mayer's tables.

Lalande's Use of Historical Evidence

Lalande's treatment of historical observations in his study of the moon's secular acceleration exhibited neither the ingenuity of Richard Dunthorne's work nor the thoroughness of Tobias Mayer's. Lalande relied heavily upon his predecessors' analyses of the ancient observations, confining himself to a brief re-evaluation of three eclipses which he thought were the most useful: the two solar eclipses

---

[24] The *Astronomie* was published in three editions, each substantially larger than the previous one: 1764 (2 volumes), 1771–1781 (4 volumes), 1792 (3 volumes including additions by Delambre).

[25] Lalande, *Astronomie* (1st edition), p. 583.

[26] Lalande, *Astronomie* (1st edition), p. 584.

[27] Lalande, *Journal d'un Voyage en Angleterre en 1763*; edited by Monod-Cassidy (1980), p. 45.

observed by Ibn Yūnus and the earliest eclipse recorded in the *Almagest*. His use of these observations was straightforward, effectively treating them as if they were modern observations. Lalande did, however, make explicit his belief in Ptolemy's unreliability as a reporter of his own and other ancient observations. Lalande's criticism was based upon the well-known problems with Ptolemy's equinox observations and Bullialdus's remarks on the different interpretations by Hipparchus and Ptolemy of some of the ancient eclipse observations, both of which had also been referred to by Mayer. But Lalande was firmer in his criticism and extrapolated from these problems into a condemnation of the whole of Ptolemy's work as a source of reliable observations, making the blanket statement that "the observations which have passed through the hands of Ptolemy being suspect" they should not be used. He also rejected the use of Ptolemy's observations of the planets in his analysis of the mean motions of Jupiter, Saturn, and Mars, relying only on observations from the *Almagest* if they had not been observed by Ptolemy, and favouring, wherever possible, the more accurate and reliable observations of Tycho and other recent observers.

A further sense of Lalande's view of ancient astronomy may be obtained from his *Astronomie*, first published in 1764 with two subsequent (and expanded) volumes published in 1771–1781 and 1792. The *Astronomie* was a popular and influential introductory text covering the whole of astronomy. Book 2 is devoted to the "origin and history of astronomy" and provides a fairly detailed study of the subject (82 pages in the first edition). Lalande divides astronomy into a mythological period before about 800 B.C. and a historical period after that time. He credits the Babylonians as the first people to have studied astronomy and gives an unusual explanation for why they first developed skill in this area: "because of the heat of the day they chose the time of night, for work, their exercises and their journeys, so that the spectacle of the stars occupied them, so to speak against their will".[28] This circumstance led the Babylonians to use astronomy in navigation, influenced their superstitions and gave rise to astrology. Apart from this, however, Lalande's account of Babylonian astronomy is fairly standard. He drew on Diodorus Siculus and Pliny for general comments on Babylonian astronomy, Ptolemy for Babylonian observations, and Sextus Empiricus for the Babylonian invention of the zodiac.

It is with his account of Greek astronomy that Lalande broke new ground. He divided Greek astronomy into an early period and a period after a "revolution of astronomy in 300 years B.C.".[29] The earlier period saw the development of very basic knowledge of the stars and constellations which can be seen in the works of Homer and Hesiod, and the philosophical consideration of the structure of the cosmos by Thales, Pythagorus, Plato, and Eudoxus. After this revolution came observers and mathematicians including Timocharis, Eratosthenes, and, most importantly, Hipparchus (whose name is even set in larger type than the names of other figures). For Lalande, Hipparchus was "the most intelligent and the most industrious astronomer of whose memory we have preserved".[30] Indeed, in Lalande's view almost all the basic discoveries within mathematical astronomy can be traced back to Hipparchus:

> Hipparchus first observed that the orbits of the planets were eccentric, (Ptol. *L. III. ch.* 2. & 2.) and their movement unequal: he wrote on this subject a particular treatise against Eudoxus & Callipus. He not only recognized the inequality of the moon, which seems to move faster in its perigee & slower in its apogee, as we will state in the VIIth book, referring to the theory of the moon, but he found as well the movement of the nodes of the moon, formed hypotheses & tables that represented the movements of the sun & the moon, and he would have done the same for the other planes, had he been able to have a fairly large number of observations.[31]

---

[28] Lalande, *Astronomie*, p. 62.

[29] Lalande, *Astronomie*, p. 94.

[30] Lalande, *Astronomie*, p. 96.

[31] Lalande, *Astronomie*, p. 97.

By contrast, Ptolemy was regarded by Lalande as a much lesser astronomer:

> Astronomers are convinced that Ptolemy was but a poor observer, that he has taken from Hipparchus & others who preceded him, all that is good in his work; I have given several similar proofs in the Mémoires de l'Académie for 1757, p. 420. (*See also Astron. Philil. p. 152. Instit. Astron. p. xxviii. Elem. d'Astron. p. 196. & 467.*). But this does nor preclude his book being extremely precious, since it is the only monument that is remaining to us of the history of astronomy & the ancient observations. In fact we can say that this book is the only one that perpetuated astronomy from Ptolemy to the time of Copernicus, for fourteen centuries of ignorance.[32]

Lalande's criticism of Ptolemy went much further than his own statements in his 1757 paper on the secular acceleration, or indeed those of any other writer. Not only was Ptolemy a poor and untrustworthy observer, who adjusted observations so that they would fit his theories, but even his theories were taken over from earlier astronomy (in particular Hipparchus). Lalande does not attempt to justify this claim—it would be hard to do so as the only detailed evidence he had for Hipparchus's theories came from Ptolemy's references to them—but it was taken up by several later historians of astronomy, in particular Lalande's student Delambre, and still has a following among certain writers today.

Lalande's appreciation of Hipparchus and his hostility towards Ptolemy, evident in his 1757 paper but more clearly stated in the *Astronomie*, certainly influenced his analysis of the secular accelerations. Lalande avoided using any of the observations made by Ptolemy when studying the acceleration of the moon and the planets, and made a fuss of rejecting them when considering the sun. One might have expected, however, that Lalande would have used Hipparchus's eclipse observations to investigate the moon's acceleration, since Lalande thinks so highly of Hipparchus. A possible explanation, which I have suggested above, is simply that Lalande relied only on those eclipses discussed by Dunthorne in his 1749 paper, and Dunthorne did not quote any of Hipparchus's observations. Instead, Lalande relied upon the most ancient Babylonian observation and the two solar eclipses observed by Ibn Yūnus, who he described in the *Astronomie* as "a careful observer from which we have three eclipses observed with precision, the only ones of all of Arabic astronomy which may be used to determine exactly the secular inequality of the moon from that time".[33]

---

[32] Lalande, *Astronomie*, p. 103.

[33] Lalande, *Astronomie*, p. 108.

Chapter 9
Epilogue

> *Yet all of Ptolemy's Almagest seems to me to breathe an air of perfect sincerity.*
>
> —Simon Newcomb, *Researches on the Motion of the Moon* (1878), p. 20.

In a little under a decade beginning in 1749 and ending in 1757, the secular acceleration of the moon had gone from being a postulated and fairly widely accepted but still unquantified phenomenon to a proven fact, whose size had been estimated three times with differing but not completely incompatible results. Dunthorne's determination of the size of the coefficient of the moon's secular equation as $10''$ per century^2 was effectively confirmed by Lalande and became the accepted figure in later eighteenth-century theoretical investigations of the secular acceleration, especially after the publication of Mayer's revised and not-too-discordant value of $9''$ per century^2 in his final lunar tables edited by Maskelyne in 1770.

As I have argued in the preceding three chapters, although Dunthorne, Mayer, and Lalande all used ancient astronomical data to investigate the moon's secular acceleration, they approached the problem in different ways. Dunthorne was certainly the most innovative. Recognizing that many of the ancient eclipse observations reported by Ptolemy were imprecisely timed, ambiguous in their terminology, and often inaccurate, Dunthorne developed a new technique whereby he could in certain cases exploit the very fact that an eclipse was seen at all to place critical constrains on the size of the secular equation. Dunthorne was also the only author to attempt to take into account the changing effects of parallax and lunar latitude on the time of maximum phase of a solar eclipse as the longitude of the moon and sun at conjunction, and therefore the time of conjunction, changes due to applying a correction for the secular equation. Mayer took a different tack, applying an iterative procedure to the problem, where he established a preliminary value for the coefficient of the secular equation and then used that value to calculate lunar positions for other ancient observations in order to refine his preliminary value. Mayer used a broader range of data than Dunthorne, including values for the mean motion of the moon from earlier astronomical tables and lunar occultations in addition to the historical eclipse records. Mayer was the first to incorporate the secular acceleration into his lunar tables. Finally, Lalande's paper was an attempt to settle the question of the size of the coefficient of the secular equation, providing what Lalande clearly hoped would be the definitive answer to the problem that would be used by other astronomers. Lalande's analysis of the secular acceleration was, however, the least impressive of the three. He relied on only three observations, and based most of his analysis of Dunthorne's earlier work. His result for the coefficient of the secular acceleration was only fractionally different to Dunthorne's, and, as we will see, it was usually to Dunthorne's study that later astronomers referred.

The existence of the moon's secular acceleration was quickly accepted following the work of Dunthorne, Mayer, and Lalande. As early as 1753, 4 years after the publication of Dunthorne's paper

J.M. Steele, *Ancient Astronomical Observations and the Study of the Moon's Motion (1691–1757)*, Sources and Studies in the History of Mathematics and Physical Sciences, DOI 10.1007/978-1-4614-2149-8_9, © Springer Science+Business Media, LLC 2012

and the same year as Mayer's study appeared, the secular acceleration was mentioned in a chronological study by George Costard. Costard was investigating the date of the eclipse supposedly foretold by Thales and explained that he had adjusted his calculation to take into account the secular
acceleration:

> You will see, Sir, how this agrees with what is said in the Petersburg Acts, pag. 332. which, therefore, I shall not
> transcribe. I shall only add, that, if any allowance is to be made for the moon's acceleration, or any other cause,
> the track here given, as you know, will be a little different. As I cannot make several ancient eclipses that I have
> tried, succeed to my mind, without some such supposition, I have done the same with regard to this. What the
> quantity to be allowed it, I leave to you and others to determine: At present I make it 45′; at Mr. Whiston's rate
> of 1′ in 54 years, or thereabouts.[1]

In a second paper that year Costard again referred to the acceleration: "But if you will make the
same allowance, as I did in my last, for the moon's acceleration, or the small retardation of the earth's
diurnal motion, the place of the center will be found at the following times …".[2] Although Costard
used the incorrect secular correction of Whiston (to my knowledge Costard was the only person other
than Whiston ever to have referred to this correction), he assumed knowledge of what the acceleration
was among his audience, which suggests that after Dunthorne's paper the existence of the acceleration was broadly accepted among the scientific community in London.

By the 1760s references to the secular acceleration became increasingly common. Lalande himself, as I have discussed in the preceding chapter, included a discussion of the secular acceleration in
his discussion of the moon's motion in his widely read *Astronomie* of 1764. The same year saw the
publication of the first part of the second volume of Roger Long's *Astronomy*, which included a four-
page discussion of the secular acceleration. Long began by describing Halley's announcement of his
discovery of the secular acceleration and the rejection of its existence by Struyck, whose arguments
Long roundly (and correctly) rejected. He then gave a detailed summary of Dunthorne's paper on the
moon's acceleration (silently correcting the typographical error in the date of the 383 B.C. eclipse
from Dunthorne's paper) before turning to the question of the cause of the acceleration:

> If the motion of the moon from the sun be accelerated, that is, if the synodical month described § 957, 958,
> appears shorter now than in ancient times, as consisting of a less number of minutes, seconds, &c, this must be
> owing to one or more of these causes; either 1, the annual and diurnal motion of the earth continuing the same,
> the moon is really carried round the earth with a greater velocity than heretofore: or 2, the diurnal motion of the
> earth and the periodical revolution of the moon continuing the same, the annual motion of the earth around the
> sun is a little retarded; which makes the sun's apparent motion in the ecliptic a little slower than formerly, and,
> consequently, the moon in passing from any conjunction with the sun spends less time before she again over-
> takes the sun, and forms a subsequent conjunction: in both these cases, the motion of the moon from the sun is
> really accelerated, and the synodical month actually shortened: or 3, the annual motion of the earth and the
> periodical revolution of the moon continuing the same, the rotation of the earth round its axis is a little retarded;
> in this case, days, minutes, seconds, &c, by which all periods of time must be measured are of a longer duration,
> and consequently the synodical month will appear to be shortened, though it really contains the same quantity
> of absolute time as it always did. When we say the moon's motion is accelerated, we would not be understood
> to determine from which of the causes now mentioned such acceleration does arise.[3]

Long gave three possible explanations for the observed secular acceleration of the moon here. First,
the moon really is speeding up. Secondly, the observed secular acceleration is really an apparent
effect caused by a slowing down in the motion of the Earth around the sun. In this case, however, we
would also observe a change in the length of the year, but Long does not address that issue. In the
next paragraph, Long speculated on possible causes of a change in the Earth's motion around the sun
or the moon's motion around the Earth:

---

[1] Costard, "Concerning the Year of the Eclipse foretold by Thales", pp. 24–25.
[2] Costard, "Concerning an Eclipse mention'd by Xenophon", pp. 158–159.
[3] Long, *Astronomy*, pp. 435–436.

> If the quantity of matter in the body of the sun be lessened by the particles of light continually streaming from it, the motion of the earth round the sun may grow slower: if the earth increases in bulk, the motion of the moon round the earth may be quickened thereby. Some are of opinion that the earth may increase in bulk by absorbing the particles of light which are continually falling upon it, or may receive an accession of matter from the tail of a comet. May not the motion of the moon round the earth have been retarded, by the near approach of a comet?[4]

This idea, that the change in motion of the Earth or the moon is caused by a change in the mass of the sun or the Earth, either by light particles being lost from the sun or light or dust particles landing on the Earth, is based upon Newton's arguments in *Principia* book 3 proposition 42.

Long's third explanation was that the observed secular acceleration might be an apparent effect caused by a change in the rate of rotation of the Earth. As our timescale is based upon the Earth's daily rotation, a change in the rate of rotation would result in a non-uniform timescale. Because we assume a uniform timescale when making calculations, the observed secular acceleration of the moon may be a consequence of the difference between the assumed uniform timescale and a true non-uniform timescale. Euler raised the issue of a change in the rate of rotation of the Earth as a possible cause of observed changes in the motion of the sun and moon in his 1750 paper "Concerning the Contraction of the Orbits of the Planets". Probably at Euler's suggestion, the Berlin Academy set as its prize essay for 1754 the question of whether the Earth's rate of rotation has changed over time and what may be the cause of such a change.[5] Although not a formal entry into the prize competition, in June 1754 Immanuel Kant published an essay on the topic in the *Wöchentliche Frag-und Anseigungs-Nachrichten*, a weekly journal published in Königsburg.[6] Kant stressed that he addressed the topic from a purely theoretical side because he believed that the historical data relating to the length of the year and intercalation "so obscure and its accounts so unreliable, as regards the question under consideration, that any theory which might be devised on that basis to make it accord with the principles of nature would probably seem to savour of the imagination".[7] Kant set out the problem as follows:

> The Earth turns unceasingly round its axis with a free motion which, having been impressed upon it once for all at the time of its formation, would continue thenceforth unaltered for all infinite time and go on with the same velocity and direction, did no impediments or external causes exist to retard or to accelerate it. I proceed to show that such an external cause actually exists, and that it is really a cause which gradually diminishes the motion of the earth and tends even to destroy its rotation, in the course of immensely long periods.[8]

Kant's explanation for this gradual slowing down of the Earth's rotation was that the constant motion of the water in the oceans from east to west caused by tides opposes Earth's rotation on its axis.[9] Although this effect is extremely small, its cumulative effect will be significant on enormous timescales. Kant's paper, however, passed largely unnoticed until the idea of a tidal effect on the Earth's rotation was revived in the nineteenth century.

The most widely accepted explanation for the secular acceleration of the moon in the 1750s and 1760s was Euler's claim of a resistance from the æther to the motion of the heavenly bodies. Bossut and D'Alembert supported this idea and Euler reaffirmed his belief in the resisting effect of the æther in his discussions of the lunar theory submitted to the Académie Royale des Sciences in 1770 and 1772.[10] Such was the interest in this matter that the Académie Royale des Sciences

---

[4] Long, *Astronomy*, pp. 435–436.

[5] Hastie (1900), p. xxxix.

[6] An English translation of Kant's essay is given by Hastie (1900), pp. 1–11. All quotations from Kant's essay are taken from Hastie's translation.

[7] Hastie (1900), p. 4.

[8] Hastie (1900), p. 4.

[9] Cartwright (1999), p. 145.

[10] Wilson (1985), p. 21.

set the topic for prize competitions in 1772 and 1774. The former was won by Euler, but in the latter year Joseph-Louis Lagrange was awarded the prize. Lagrange's essay was titled "Sur l'Équation Séculaire de la Lune" and was published in the *Mémoires de Mathématique et de Physique, Présentés à l'Académie Royale des Sciences, par divers Savans, & Iûs dans ses Assemblées* for 1773 (which actually appeared in 1776). In this paper Lagrange investigated whether the secular acceleration could be caused by the gravitational effects of the non-spherical shape of the Earth, concluding that it could not (he later showed that the non-spherical shape of the moon also could not explain the acceleration). Lagrange also reviewed the evidence for the existence of the secular acceleration given by Dunthorne, Mayer, and Lalande, reviewing Dunthorne's paper in particular in some detail.

Probably inspired by Euler's work and the prize competitions, other astronomers were also drawn to the question of the moon's secular acceleration around that time. In the same volume of *Mémoires de Mathématique et de Physique, Présentés à l'Académie Royale des Sciences, par divers Savans, & Iûs dans ses Assemblées* as Lagrange's prizewinning essay, Pierre-Simon Laplace published "Sur le principle de la Gravitation universelle, & sur les Inégalités Séculaires des Planètes qui en dépendent" in which he argued that the action of gravity acts with a finite speed and its action depends upon the speed and direction of motion of a body which it acts upon, which, he claimed, would result in an acceleration in the motion of the moon and planets.[11] Laplace referred to Mayer for information on the observed secular acceleration.

That same year, Jean Bernoulli, inspired by the 1770 publication of Mayer's improved lunar tables, undertook a critical review of Dunthorne's 1749 paper.[12] Using Mayer's new lunar tables and Lacaille's solar tables, Bernoulli calculated the circumstances of the eclipses studied by Dunthorne (noting the misprints in Dunthorne's paper). Although he did not deny the existence of the secular acceleration, Bernoulli urged caution: "The times of the most ancient eclipses, & even those which are less, appear to me to be too vaguely indicated and to be too uncertain that they could be adopted with confidence, & in particular as we want to use the error of tables to determine the size of the secular equation".[13] Bernoulli relied only on those ancient observations discussed by Dunthorne in forming this judgement, however; he did not consider the other eclipses in the *Almagest*, for example.

By the end of the eighteenth century a theoretical derivation of the moon's secular acceleration from the law of gravitation had been formulated by Laplace which predicted a coefficient of the secular equation of a little over $10''$ per century^2, in good agreement with the coefficient Dunthorne had found from the ancient observations. The problem of the moon's secular acceleration appeared to be solved and attention turned away from attempts to determine the size of the acceleration from ancient observations. A notable exception was an unusual and largely forgotten treatise entitled *Erreur des Astronomes et des Gómetres d'avoir admis l'Accélération Séculaire de la Lune*, published by a Jean-Baptiste-Phillipe Marcoz in 1833. Marcoz made the bold assertion that astronomers had been misled by the ancient records and that the secular acceleration did not exist. Furthermore, foreshadowing Robert Newton's *The Crime of Claudius Ptolemy* (1977) by a century and a half, Marcoz claimed that all the eclipse reports in the *Almagest* had been faked by Ptolemy. Without these reports, he said, there was no empirical evidence for the secular acceleration, and therefore Laplace's derivation of a theoretical cause for the acceleration was baseless. Marcoz sent his book to all the major scientific institutions in France and England, but his work seems to have been dismissed by other astronomers and quickly forgotten.

---

[11] Wilson (1985), p. 21.

[12] Bernoulli, "Mémoire sur la Comparaison de quelques Observations anciennes de la Lune acec les Tables de Mayer".

[13] Bernoulli, "Mémoire sur la Comparaison de quelques Observations anciennes de la Lune acec les Tables de Mayer", p. 179.

The discovery of a mistake in Laplace's model by John Couch Adams in the early 1850s reopened the question of the moon's secular acceleration.[14] Ptolemy's observations again became a matter of debate between astronomers, some astronomers placing their trust in them, others arguing for the use of other observations, in particular, untimed solar eclipses reported in classical histories. The eclipse of 23 December 383 B.C. caused particular concern. Dunthorne had used the visibility (but not the timing) of this eclipse to constrain his value for the secular acceleration, but the eclipse had been discarded as problematical by Mayer. But as lunar and solar theories improved during the nineteenth century it became clear that this eclipse was incompatible with the secular acceleration obtained from all of the other eclipses reported in the *Almagest*. Simon Newcomb, who was generally very supportive of the reliability of the *Almagest* eclipse records, concluded that the observation of eclipse must have been a mistake.[15] Other authors took a harsher view, using this record as evidence of Ptolemy's unreliability.[16]

By the end of the nineteenth and through the twentieth century, the *Almagest* eclipse records started to lose their importance for the study of the moon's motion. The full collection of observations preserved in Golius' manuscript of Ibn Yūnus was eventually made available and the observations were used by Newcomb in his work on the secular acceleration. Then, in the second half of the twentieth century, F. Richard Stephenson and others utilized many hundreds of Chinese and Babylonian observations in their work (now phrased in terms of the long-term change in the Earth's rate of rotation following the independent establishment of the lunar acceleration using occultations of Mercury and, more recently, lunar laser ranging).[17]

Interest in ancient astronomical records over the past 200 years has not been confined to astronomers studying the moon's secular acceleration, however. The nineteenth century also saw an upturn in the study of chronology and the history of astronomy. In 1806 Ludwig Ideler published *Historische Untersuchungen über die astronomischen Beobachtungen der Alten*, the first serious attempt to study ancient astronomical observations and their application to chronology since the mid-seventeenth century.[18] A decade later, Jean Baptiste Joseph Delambre published the two-volume *Histoire de l'Astronomie Ancienne*, the first truly historical attempt to understand ancient astronomy. Through the research for this book, Delembre became the first person to understand Ptolemy in detail since the seventeenth century. Like his teacher Lalande, Delambre was deeply sceptical of Ptolemy's contributions as an astronomer.

The growth of studies such as those of Ideler and Delambre during the nineteenth century reflects the beginning of a new phase in the relationship between scholars and ancient astronomy. For a century and a half, ancient astronomy had become forgotten territory, no longer the living tradition it had been during the Renaissance and into the beginning of the seventeenth century. Astronomers and historians no longer understood the details of Ptolemy's methods, no longer relied upon his observations except in a few instances, and no longer saw themselves as building upon an ancient foundation. With the work of Delambre and Ideler at the beginning of the nineteenth century, ancient astronomy once more became relevant, and, crucially, an understanding of its technical details was recovered. This rediscovery of ancient astronomy was not aimed at relearning ancient methods that could be incorporated into current astronomy, but instead, for the first time, ancient astronomy became a topic

---

[14] On the subsequent history of the theoretical investigation of the moon's secular acceleration, see Kushner (1989), Britton (1992), pp. 153–178, and Wilson (2010), pp. 239–284.

[15] Newcomb, *Researches on the Motion of the Moon*, p. 43.

[16] The recent discovery of an account of this eclipse preserved on a cuneiform tablet from Babylon has cleared Ptolemy of the charge of misrepresenting the Babylonian account of the observation. See Steele (2005).

[17] Stephenson (1997).

[18] Ideler followed this work with his monumental two-volume *Handbuch der mathematischen und technischen Chronologie* in 1825–1826.

primarily of historical interest. Scientific techniques were no longer learnt by reading the texts of ancient astronomy as they had been before the eighteenth century but were now used to understand the history of that ancient astronomy. And, as a greater understanding of the history of ancient astronomy was developed, the opportunity arose (sadly not always seized, even today) for ancient observations to be used more judiciously and effectively in solving astronomical problems where they provide only the empirical data.

References

Primary Sources

Manuscripts

University Library, Cambridge
MS RGO 2.6
MS RGO 2.8
MS RGO 2.16
MS RGO 4.125

Niedersächische Staats- und Universitäts-Bibliothek, Göttingen
MS Mayer 9
MS Mayer 15_6
MS Mayer 15_{28}
MS Mayer 15_{33}
MS Mayer 15_{41}
MS Mayer 15_{48}

The British Library, London
BL Slone MS 2281
BL MS Add. 4224
BL MS Add. 5489

Museum of the History of Science, Oxford
MS Museum 95

Printed Works

[Anon.], 1733, *Histoire de l'Academie Royale des Sciences Depius 1686 jusqu'a son Renouvellement en 1699* (Paris).

[Anon.], 1733, *Phlegon's Testimony Shewn to Relate to the Darkness Which happened at our Saviour's Passion. In a Letter to Dr. Sykes* (London).

B. J. [John Bevis], 1754, 'Mayer's new Tables of the Sun and Moon', *The Gentleman's Magazine* 24, 374–376.

Barrettus, Lucius [Albertus Curtius], 1666, *Historia Coelestis ex libris commentariis manuscriptis observationum vicennalium viri generosi Tichonis Brahe Dani* (Augsberg).

Bernoulli, Jean, 1775, 'Mémoire sue la Comparaison de quelques Observations anciennes de la Lune avec les Tables de Mayer', *Nouveaux Mémoires de l'Académie Royale des Sciences et Belles-Lettres. Année MDCCLXXIII*, 177–191.

Brahe, Tycho, 1598, *Astronomiæ Instauratæ Mechanica* (Wandesburg).

Brahe, Tycho, 1602, *Astronomiæ Instauratæ Progymnasmatum* (Prague).

J.M. Steele, *Ancient Astronomical Observations and the Study of the Moon's Motion (1691–1757)*, 143
Sources and Studies in the History of Mathematics and Physical Sciences,
DOI 10.1007/978-1-4614-2149-8, © Springer Science+Business Media, LLC 2012

Brent, Charles, 1741, *The Compendious Astronomer: Containing New and Correct Tables For Computing in a concise Manner, the Places of the Luminaries; Digested from Numbers Founded on The latest Observations* (London).

Bullialdus, Ismael, 1645, *Astronomia Philolaica* (Paris).

Cassini, Jacques, 1740, *Éléments d'Astronomie* (Paris).

Cassini, Jacques, 1740, *Tables Astronomiques, du Soleil, de la Lune, des Planetes, des Étoiles fixes, et des satellites de Jupiter et de Saturne, avec l'Explication et l'Usage de ces mêmes tables* (Paris).

Chapman, John, 1734, *Phlegon Examined Critically and Impartially. In Answer to the late Dissertation and Defence of Dr. Sykes. To which is added a Postscript, Explaining a Passage in Tertullian* (Cambridge).

Chapman, John, 1735, *Phlegon Re-examined: In Answer to Dr. Sykes's Second Defence of his Dissertation Concerning Phlegon. To which is added a Postscript, Concerning the Chronicon Paschale* (Cambridge).

Clairaut, Alexis, 1754, *Tables de la Lune, Calculées suivant la Théorie de la Gravitation Universelle* (Paris).

Clarke, Samuel, 1706, *A Discourse Concerning the Unchangeable Obligations of Natural Religion, and the Truth and Certainty of the Christian Revelation* (London).

Costard, George, 1746, *A Letter to Martin Folkes, Esq., President of the Royal Society, Concerning The Rise and Progress of Astronomy amongst the Ancients* (London).

Costard, George, 1746, 'A Letter from the Rev. Mr. G Costard, to the Rev. Thomas Shaw, D. D. F. R. S. and Principle of St. Edmunds-Hall, concerning the Chinese Chronology and Astronomy', *Philosophical Transactions of the Royal Society* 44, 476–493.

Costard, George, 1748, *A Further Account of the Rise and Progress of Astronomy amongst the Ancients, in Three Letters to Martin Folkes, Esq; President of the Royal Society* (London).

Costard, George, 1753 'A Letter from the Rev. Mr. George Costard, Fellow of Wadham-College, Oxford, to Dr. Bevis, concerning the Year of the Eclipse foretold by Thales', *Philosophical Transactions of the Royal Society* 48, 17–26.

Costard, George, 1753, 'A Letter from the Rev. Mr. George Costard to Dr. Bevus, concerning an Eclipse mention'd by Xenophon', *Philosophical Transactions of the Royal Society* 48, 155–160.

Costard, George, 1764, *The Use of Astronomy in History and Chronology Exemplified in An Inquiry into the Fall of the Stone into the Ægospotamos* (London).

Costard, George, 1767, *The History of Astronomy, With Its Application to Geography, History and Chronology, Occasionally Exemplified by Globes* (London).

Costard, George, 1777, 'Translation of a Passage in Ebn Younes: With Some Remarks Thereon: In a Letter from the Rev. George Costard, M. A. Vicar of Twickenham, to the Rev. Samuel Horsley, LL. D. Sec. R. S.', *Philosophical Transactions of the Royal Society of London* 67, 231–243.

Delambre, Jean Baptiste Joseph, 1817–1818, *Histoire de l'Astronomie Ancienne* (2 volumes, Paris).

Disney, John, 1785, *Memoires of the Life and Writing of Arthur Ashley Sykes, D.D.* (London).

Du Halde, Jean-Baptiste, 1736, *Description Géographique, Historique, Chronologique, Politique, et Physique de l'Empire de la Chine et de la Tartarie Chinoise, Enrichie des Cartes Générales et Particulieres de ces Pays, de la Carte générale et des Cartes particulieres du Thibet, & de la Corée; & ornée d'un grand nombre de Figures & de Vignettes gravées en Taille-douce* (4 volumes, Paris).

Du Halde, Jean-Baptiste, 1736, *The General History of China. Containing a Geographical, Historical, Chronological, Political and Physical Description of the Empire of China, Chinese-Tartary, Corea and Thibet. Including an Exact and Particular Account of their Customs, Manners, Ceremonies, Religion, Arts and Sciences* (4 volumes, London).

Dunthorne, Richard, 1739, *The Practical Astronomy of the Moon: or, New Tables of the Moon's Motions, Exactly Constructed from Sir Isaac Newton's Theory, as Published by Dr Gregory in his Astronomy* (Cambridge).

Dunthorne, Richard, 1747, 'A Letter from Mr. Richard Dunthorne, to the Rev. Mr. Cha. Mason, F. R. S. and Woodwardian Professor of Nat. Hist. at Cambridge, concerning the Moon's motion', *Philosophical Transactions of the Royal Society* 44, 412–420.

Dunthorne, Richard, 1749, 'A Letter from the Rev. Mr. Richard Dunthorne to the Reverend Mr. Richard Mason F. R. S. and Keeper of the Woodwardian Museum at Cambridge, concerning the Acceleration of the Moon', *Philosophical Transactions of the Royal Society* 46, 162–172.

Dunthorne, Richard, 1751, 'A Letter from Mr. Rich. Dunthorne to the Rev. Dr. Long, F. R. S. Master of Pembroke-Hall in Cambridge, and Lowndes's Professor of Astronomy and Geometry in that University, concerning Comets', *Philosophical Transactions of the Royal Society* 47, 281–288.

Estève, Pierre, 1755, *Histoire Generale et Particuliere de l'Astronomie* (3 volumes, Paris).

Euler, Leonard, 1746, *Opuscula varii argumenti* (Volume 1, Berlin).

Euler, Leonard, 1749, 'Part of a Letter from Leonard Euler, Prof. Math. at Berlin, and F. R. S. To the Rev. Mr. Caspar Wetstein, Chaplain to his Royal Highness the Prince of Wales, concerning the gradual Approach of the Earth to the Sun', *Philosophical Transactions of the Royal Society* 46, 203–305.

Euler, Leonard, 1750, 'Part of a Letter from Mr. Professor Euler To the Reverend. Mr. Wetstein, Chaplain to his Royal Highness the Prince, concerning the Contractions of the Orbits of the Planets', *Philosophical Transactions of the Royal Society* 46, 356–359.

Ferguson, James, 1794, *Astronomy Explained upon Sir Isaac Newton's Principals* (9th edition, London).

Flamsteed, John, 1725, *Historia Coelestis Britannica* (3 volumes, London).

Gregory, David, 1702, *Astronomiæ Physicæ & Geometriæ Elementa* (London).

Gregory, David, 1715, *The Elements of Astronomy, Physical and Geometrical* (London).

Halley, Edmond, 1686–92, 'Emendariones & Notæ in tria vitiose edita in Textu vulgato Naturalis Historiæ C. Plinni', *Philosophical Transactions of the Royal Society* 16, 535–540.

Halley, Edmond, 1986–92, 'A Discourse tending to prove at what Time and Place, Julius Cesar made his first Descent upon Britain', *Philosophical Transactions of the Royal Society* 16, 495–501.

Halley, Edmond, 1686–92, 'An Account of the Cause of the Change of the Variation of the Magnetical Needle; With an Hypothesis of the Structure of the Internal Parts of the Earth; As it Was Proposed to the Royal Society in One of Their Late Meetings', *Philosophical Transactions of the Royal Society* 16, 563–578.

Halley, Edmond, 1963, 'Emendationes ac Notæ in vetustas *Albatênii* Observationes Astronomicas, cum restitione Tabularum Lunisolarium eiusdem Authoris', *Philosophical Transactions of the Royal Society* 17, 913–921.

Halley, Edmond, 1695 'Some Account of the Ancient State of the City of Palmyra, with short Remarks upon the Inscriptions found there', *Philosophical Transactions of the Royal Society* 19, 160–175.

Halley, Edmond, 1749, *Tabulæ Astronomicæ Accedunt De Usu Tabularum Præcepta* (London).

Halley, Edmond, 1752, *Astronomical Tables with Precepts, both in English and Latin, for Computing Places of the Sun, Moon, Planets, and Comets* (London).

Heathcote, Ralph, 1747 *Historia Astronomiæ, sive, De Ortu & Progressu Astronomiæ* (Cambridge).

Horrocks, Jeremiah, 1678, *Opera Posthuma* (London).

Ideler, Ludwig, 1806, *Historische Untersuchungen über die astronomischen Beobachtungen der Alten* (Berlin).

Ideler, Ludwig, 1825–1826, *Handbuch der mathematischen und technischen Chronologie* (2 volumes, Berlin).

Keill, John, 1721, *An Introduction to the True Astronomy: or, Astronomical Lectures Read in the Astronomical School at the University of Oxford* (London).

Kepler, Johannes, 1615, *Eclogæ Chronicæ* (Frankfurt).

Kepler, Johannes, 1618–1621, *Epitome Astronomiæ Copernicanæ* (Frankfurt).

Kepler, Johannes, 1627, *Tabule Rulolphinæ, Quibus Astronomicæ Scientiæ, Temporum longinquitate collapsæ Restauration continetur* (Ulm).

Lagrange, Joseph-Louis, 1776, 'Sur l'Équation Séculaire de la Lune', *Mémoires de Mathématique et de Physique, Présentés à l'Académie Royale des Sciences, par divers Savans, & Iûs dans ses Assemblées. Année 1773*, 1–61.

Lalande, Jérôme, 1757, 'Mémoire sur les Équations Séculaires, Et sur les moyens mouvemens du Soleil, de la Lune, de Saturne, de Jupiter, & de Mars, Avec les observations de Tycho-brahé, faites sur Mars en 1593, tirées des manuscripts de cet Auteur', *Mémoires de l'Académie Royale des Sciences*, 441–470.

Lalande, Jérôme, 1763, *Journal d'un Voyage en Angleterre en 1763* (unpublished). [= Monod-Cassidy (1980)].

Lalande, Jérôme, 1764, *Astronomie* (1st edition, Paris).

Laplace, Pierre-Simon, 1776, 'Sur le principle de la Gravitation universelle, & sure les Inégalités Séculaires des Planètes qui en dépendent', *Mémoires de Mathématique et de Physique, Présentés à l'Académie Royale des Sciences, par divers Savans, & Iûs dans ses Assemblées. Année 1773*, 163–232.

Leadbetter, Charles, 1727, *Astronomy; or, The True System of the Planets Demonstrated* (London).

Leadbetter, Charles, 1735, *Uranoscopia: or, The Contemplation of the Heavens* (London).

Leadbetter, Charles, 1742, *A Compleat System of Astronomy* (London).

Long, Roger, 1742–84, *Astronomy in Five Books* (Cambridge).

Longomontanus, Christian, 1622, *Astronomia Danica* (Amsterdam).

Marcoz, J.-B.-P., 1833, *Erreur des Astronomes et des Gómetres d'avoir admis l'Accélération Séculaire de la Lune* (Paris).

Mayer, Tobias, 1745, *Mathematischer Atlas, in welchem aus 60 Tabellen alle Theile der Mathematik vorgestellet, und nicht allein zu bequemer Wiederholung, sondern auch den Anfängern besonders zur Aufmunterung durch deutliche Beschreibung und Figuren entrofen werden* (Augsburg).

Mayer, Tobias, 1753, 'Novae Tabulae Motuum Solis et Lunae', *Commentarii Societatis Regiae Scientiarum Gottingensis* 2, 383–430.

Mayer, Tobias, 1767, *Theoria Lunæ Juxta Systema Newtonianum* (London).

Mayer, Tobias, 1770, *Tabulæ Motuum Solis et Lunæ Novæ et Correctæ; Auctore Tobia Mayer: Quibus Accedit Methodus Longitudinum Promota, Eodem Auctore* (London).

Montucla, Jean Etienne, 1758, *Histoire des Mathematiques* (1st edition: 2 volumes, Paris).

Newcomb, Simon, 1878, *Researches on the Motion of the Moon made at the United States Naval Observatory, Washington. Part I. Reduction and Discussion of Observations of the Moon Before 1750* (Washington).

Newton, Isaac, 1687, *Philosophiæ Naturalis Principia Mathematica* (1st edition, London).

Newton, Isaac, 1702, *Theory of the Moon's Motion* (London).

Newton, Isaac, 1713, *Philosophiæ Naturalis Principia Mathematica* (2nd edition, London).

Newton, Isaac, 1726, *Philosophiæ Naturalis Principia Mathematica* (3rd edition, London).

Riccioli, Giovanni Battista, 1651, *Almagestum Novum* (Bolognia).

Riccioli, Giovanni Battista, 1665, *Astronomiæ Reformatæ* (Bolognia).

Scriblerus, Simon, 1731, *Whistoneutes: or, Remarks on Mr. Whiston's Historical Memoires of the Life of Dr. Samuel Clarke, &c* (London).

Smith, George, 1748, *A Dissertation on the General Properties of Eclipses; And particularly the ensuing Eclipse of 1748, considered thro' all its Periods* (London).

Souciet, Etienne, 1729–32, *Observations Mathématiques, Astronomiques, Géorgraphiques, Chronologiques, et Physiques; Tirées des Anciens Livres Chinois, ou faites nouvellement aux Indes, à la Chine & aileurs, par les Pères de la Compagnie des Jesus* (3 volumes, Paris).

Streete, Thomas, 1661, *Astronomia Carolina. A New Theorie of the Cœlestial Motions* (1st edition, London).

Streete, Thomas, 1710, *Astronomia Carolina. A New Theorie of the Cœlestial Motions* (2nd edition, London).

Streete, Thomas, 1716, *Astronomia Carolina. A New Theorie of the Cœlestial Motions* (3rd edition, London).

Struyck, Nicolaas, 1740, *Inleiding tot de Algemeene Geographie, benevens eenige Sterrekundige en andere Verhandelingen* (Amsterdam).

Sykes, Arthur Ashley, 1732, *A Dissertation on the Eclipse Mentioned by Phlegon. Or, An Enquiry Whether That Eclipse Had Any Relation to the Darkness Which Happened at Our Saviour's Passion* (London).

Sykes, Arthur Ashley, 1733, *A Defence of the Dissertation on the Eclipse Mentioned by Phlegon: Wherein it is further shewn, That the Eclipse had no Relation to the Darkness which happened at our Saviour's Passion: And Mr. Whiston's Observations are particularly considered* (London).

Sykes, Arthur Ashley, 1734, *A Second Defence of the Dissertation upon the Eclipse mentioned by Phlegon: Wherein Mr. Chapman's Objections, and Those of the A. of the Letter to Dr. Sykes, are particularly considered* (London).

Weidler, Johannes Frederic, 1741, *Historia Astronomiae sive de Ortu et Progressu Astronomiae* (Vitemberg).

Whalley, John, 1701, *Ptolemy's Quadupartite; or, Four Books Concerning The Influences of the Stars* (London).

Whiston, William, 1715, *Astronomical Lectures Read in the Publick Schools at Cambridge; Whereunto is added a Collection of Astronomical Tables; Being those of Mr. Flamsteed, Corrected; Dr. Halley; Monsieur Cassini; and Mr. Street* (London).

Whiston, William, 1715, *The Copernicus Explained. Or a Brief Account of the Nature and Use of an Universal Astronomical Instrument for the Calculation and Exhibition of New and Full Moons, and of Eclipses, both Solar and Lunar; with the Places Heliocentrical and Geocentrical of all the Planets, Primary and Secondary, &c.* (London).

Whiston, William, 1730, *Historical Memoirs of the Life of Dr. Samuel Clarke. Being a Supplement to Dr. Sykes's and Bishop Hoadley's Accounts. Including certain Memoires of several of Dr. Clarke's Friends* (London).

Whiston, William, 1732, *The Testimony of Phlegon Vindicated: Or, An Account of the great Darkness and Earthquake at our Savior's Passion, described by Phlegon* (London).

Whiston, William, 1734, *Six Dissertations* (London).

Whiston, William, 1749, *Memoirs of the Life and Writings of Mr. William Whiston containing Memoirs of several of his Friends also* (London).

Wing, Vincent, 1652, *Astronomia Britannica* (London).

Wing, Vincent, 1656, *Astronomia Instaurata: Or, A new and compendious Restauration of Astronomie in Four Parts* (London).

Wright, Robert, 1732, *New and Correct Tables of the Lunar Motions, According to the Newtonian Theory* (Manchester).

Secondary Sources

Aiton, E. J. (ed.), 1996, *Leonhardi Euleri Opera Omnia, Ser. Seunda, XXXI: Commentationes Mechanicae et Astronomicae ad Physicam Cosmicam Pertinentes* (Basil: Birkhauser).

Albrecht, S., 1999, 'Der Turm zu Babel als bildlicher Mythos: Malerei – Graphik – Architektur', in J. Renger (ed.), *Babylon: Focus Mesopotamischer Geschichte, Wiege Früher Gelehrsamkeit, Mythos in der Moderne* (Berlin: Deutsche Orient-Gesellschaft), 443–574.

Albree, J., and Brown, S. H., 2008, ' " A Valuable Monument of Mathematical Genius": The Ladies' Diary (1704–1840)', *Historia Mathematica* 36, 10–47.

Amiable, L., 1889, *Le Franc-Maçon Jérôme Lalande* (Paris: Charavay Frères).

Armitage, A., 1966, *Edmond Halley* (London and Edinburg: Thomas Nelson and Sons).

Asher-Greve, J. M., 2006, 'From 'Semiramis of Babylon' to 'Semiramis of Hammersmith", in S. W. Holloway (ed.), *Orientalism, Assyriology and the Bible* (Sheffield: Sheffield Phoenix Press), 322–373.

Attwater, A., 1936, *Pembroke College Cambridge: A Short History* (Cambridge: Cambridge University Press).

Birkett, K., and Oldroyd, D., 1991, 'Robert Hooke, Physico-Mythology, Knowledge of the World of the Ancients and Knowledge of the Ancient World', in S. Gaukroger (ed.), *The Uses of Antiquity* (Dordrecht: Kluwer), 145–170.

Britton, J. P., 1992, *Models and Precision: The Quality of Ptolemy's Observations and Parameters* (New York: Garland).

Buchwald, J. Z., and Feingold, M., forthcoming, *Reckoning with the Past: Isaac Newton, Ancient Chronicles and the Temper of Evidence*.

Burke, P., 1997, *Varieties of Cultural History* (Ithaca: Cornell University Press).

Carmody, F. J., 1956, *Arabic Astronomical and Astrological Sciences in Latin Translation: A Critical Bibliography* (Berkely and Los Angeles: University of California Press).

Cartwright, D. E., 1999, *Tides: A Scientific History* (Cambridge: Cambridge University Press).

Chapin, S. L., 1988, 'Lalande and the Length of the Year; Or, How to Win a Prize and Double Publish', *Annals of Science* 48, 183–190.

Chapman, A., 1994, 'Edmond Halley's Use of Historical Evidence in the Advancement of Science', *Notes and Records of the Royal Society* 48, 167–191.

Chapman, A., and Johnson, A. D., 1982, *The Preface to John Flamsteed's Historia Coelestis Britannica or British Catalogue of the Heavens (1725)* (Greenwich: National Maritime Museum).

Christianson, J. R., 2000, *On Tycho's Island: Tycho Brahe and His Assistants 1570–1601* (Cambridge: Cambridge University Press).

Clerke, A. M., and McConnell, A., 2004, 'Richard Dunthorne (1711–1775)', *Oxford Dictionary of National Biography* (Oxford: Oxford University Press).

Cohen, I. B., 1975, *Isaac Newton's Theory of the Moon's Motion (1702) With a Bibliographical and Historical Introduction* (Folkestone: Dawson).

Cohen, I. B., and Whitman, A., 1999, *Isaac Newton: The Principia. A New Translation and Guide* (Berkeley, Los Angeles, London: University of California Press).

Cook, A., 1998, *Edmond Halley: Charting the Heavens and the Seas* (Oxford: Clarendon Press).

Cooper, C. H., 1863, 'On Richard Dunthorne, Astronomer, Engineer, and Antiquarian Artist', *Antiquarian Communications of the Cambridge Antiquarian Society* 2, 331–335.

Crépel, P., and Coste, A., 2005, 'Jean-Etienne Montucla, *Histoire des Mathématiques*, Second Edition (1799–1802)', in I. Grattan-Guinness (ed.), *Landmark Writings in Western Mathematics, 1640–1940* (Amsterdam: Elsevier), 292–302.

Croarken, M., 2007, 'Dunthorne, Richard', in T. Hockey (ed.), *The Biographical Encyclopedia of Astronomers* (Berlin: Springer), 319.

Dahm, J. J., 1970, 'Science and Apologetics in the Early Boyle Lectures', *Church History* 39, 172–186.

Delambre, J.-B., 1827, *Histoire de l'Astronomie au Dix-Huitième Siècle* (Paris: Bachelier).

D'Elia, P. M., 1960, *Galileo in China* (Cambridge, MA: Harvard Unversity Press).

Depuydt, L., 1995, 'More Valuable Than All Gold: Ptolemy's Royal Canon and Babylonian Chronology', *Journal of Cuneiform Studies* 47, 97–117.

Derome, R., 2000–01, 'Iconography of Ptolemy's Portrait', http://www.er.uqam.ca/nobel/r14310/Ptolemy/index.html.

Dreyer, J. L. W., 1913–29, *Tychonis Brahe Dani Opera Omnia* (Copenhagen: Libraria Gyldendaliana).

Dumont, S., 2007, *Un Astronome des Lumières: Jérôme Lalande* (Paris: l'Observatoire de Paris & Vuibert).

Farrell, M., 1981, *William Whiston* (New York: Arno Press).

Fellmann, E. A., 2007, *Leonhard Euler* (Basel-Boston-Berlin: Birkhäuser Verlag).

Feisenberger, H. A. (ed.), 1975, *Sale Catalogues of Libraries of Eminent Persons. Vol. 11: Scientists* (London: Mansell).

Finocchiaro, M., 2007, *Retrying Galileo, 1633–1992* (Berkeley and Los Angeles: University of California Press).

Forbes, E. G., 1966, 'Tobias Mayer's Lunar Tables', *Annals of Science* 22, 105–116.

Forbes, E. G., 1967, 'The Life and Work of Tobias Mayer (1723–62)', *Quarterly Journal of the Royal Astronomical Society* 8, 227–251.

Forbes, E. G., 1971, *The Euler-Mayer Correspondence (1751–1755)* (New York: Elsevier).

Forbes, E. G., 1972, *The Unpublished Writings of Tobias Mayer. Vol. I: Astronomy and Geography* (Göttingen: Vandenhoeck & Ruprecht).

Forbes, E. G., 1980, *Tobias Mayer (1723–62): Pioneer of Enlightened Science in Germany* (Göttingen: Vandenhoeck & Ruprecht).

Forbes, E. G., 1983, 'La Correspondance Astronomique entre Joseph-Nicolas Delisle et Tobias Mayer', *Revue d'Histoire des Sciences* 36, 113–151.

Forbes, E. G., and Wilson, C., 1995, 'The Solar Tables of Lacaille and the Lunar Tables of Mayer', in R. Taton and C. Wilson (eds.), *The General History of Astronomy. Volume 2: Planetary Astronomy from the Renaissance to the Rise of Astrophysics. Part B The Eighteenth and Nineteenth Centuries* (Cambridge: Cambridge University Press), 55–68.

Force, J. E., 1985, *William Whiston: Honest Newtonian* (Cambridge: Cambridge University Press).

Fotheringham, J. K., 1920, 'A Solution of the Ancient Eclipses of the Sun', *Monthly Notices of the Royal Astronomical Society* 81, 104–126.

Fuchs, R., 2009, 'Die «Portraits» des Klaudios Ptolemaios', in A. Stückelberger und F. Mittenhuber (eds.), *Klaudios Ptolemaios. Handbuch der Geographie: Ergänzungsband* (Basel: Schwabe Verlag), 402–429.

Gapaillard, J, 2010, 'Lalande à Berlin et sa correspondance avec Euler', in G. Boistel, J. Lamy and C. Le Ley (eds.), *Jérôme Lalande (1732–1807): Une Trajectoire Scientifique* (Rennes: Presses Universitaires de Rennes), 87–107.

Gascoigne, J., 1989, *Cambridge in the Age of the Enlightenment* (Cambridge: Cambridge University Press).

Gaythorpe, S. B., 1957, 'Jeremiah Horrocks and his "New Theory of the Moon"', *Journal of the British Astronomical Association* 67, 134–144.

Goulding, R., 2006, 'Histories of Science in Early Modern Europe: Introduction', *Journal for the History of Ideas* 67, 33–40.

Goulding, R., 2010, *Defending Hypatia: Ramus, Savile, and the Renaissance Rediscovery of Mathematical History* (Dordrecht: Springer).

Grafton, A., 1993, *Joseph Scaliger: A Study in the History of Classical Scholarship II: Historical Chronology* (Oxford: Clarendon Press).

Grafton, A., 1997, 'From Apotheosis to Analysis: Some Late Renaissance Histories of Classical Astronomy', in D. R. Kelley (ed.), *History and the Disciplines: The Reclassification of Knowledge in Early Modern Europe* (Rochester: The University of Rochester Press), 261–276.

Grafton, A., 2003, 'Some Uses of Eclipses in Early Modern Chronology', *Journal of the History of Ideas* 64, 213–229.

Grant, R., 1852, *History of Physical Astronomy from the Earliest Ages to the Middle of the Nineteenth Century* (London: Henry G. Bohn).

Gunther, R. T., 1937, *Early Science in Cambridge* (Oxford: Oxford University Press).

Hald, A., 2003, *A History of Probability and Statistics and Their Applications before 1750* (Hoboken: John Wiley & Sons).

Hastie, W., 1900, *Kant's Cosmogony as in his Essay on the Retardation of the Rotation of the Earth and his Natural History and Theory of the Heavens* (Glasgow: James Maclehose and Sons).

Hotton, C., Shaw, G. and Pearson, R., 1809, *The Philosophical Transactions of the Royal Society of London, From their Commencement, in 1665, to the Year 1880; Abridged, with Notes and Biographic Illustrations* (London: Baldwin).

Hsia, F. C., 2008, 'Chinese Astronomy for the Early Modern European Reader', *Early Science and Medicine* 13, 417–450.

Hsia, F. C., 2009, *Sojourners in a Strange Land: Jesuits and Their Scientific Missions in Late Imperial China* (Chicago: University of Chicago Press).

Johnson, F. R., 1937, *Astronomical Thought in Renaissance England: A Study of the English Scientific Writings from 1500 to 1645* (Baltimore: The Johns Hopkins Press).

Jones, A., 2005, 'In order that we should not ourselves appear to be adjusting our estimates … to make them fit some predetermined account', in J. Z. Buchwald and A. Franklin (ed.), *Wrong for the Right Reasons* (Dordrecht: Springer), 17–39.

Jones, A., 2006, 'Ptolemy's Ancient Planetary Observations', *Annals of Science* 63, 255–290.

Jones, A., 2012, 'Theon of Alexandria's Observation of the Solar Eclipse of A.D. 364 June 16', *Journal for the History of Astronomy* 43, 117–118.

Jungnickel, C. and McCormmach, R., 1996, *Cavendish* (Philadelphia: American Philosophical Society).

Kelly, J. T., 1991, *Practical Astronomy During the Seventeenth Century: Almanac Makers in America and England* (New York: Garland).

Kollerstrom, N., 1992, 'The Hollow World of Edmond Halley', *Journal for the History of Astronomy* 23, 185–192.

Kollerstrom, N., 2000, *Newton's Forgotten Lunar Theory: His Contribution to the Quest for Longitude* (Santa Fe: Green Lion Press).

Kremer, R. L., 1981, 'The Use of Bernard Walther's Astronomical Observations: Theory and Observation in Early Modern Astronomy', *Journal for the History of Astronomy* 12, 124–132.

Kushner, D., 1989, 'The Controversy Surrounding the Secular Acceleration of the Moon's Mean Motion', *Archive for History of Exact Sciences* 39, 291–316.

Laudan, R., 1993, 'Histories of the Sciences and their Uses: A Review to 1913', *History of Science* 31, 1–33.

Linton, C. M., 2004, *From Eudoxus to Einstein: A History of Mathematical Astronomy* (Cambridge: Cambridge University Press).

Llewellyn-Jones, L., and Robson, J., 2010, *Ctesias' History of Persia: Tales of the Orient* (Abingdon: Routledge).

Lundquist, J. M., 1995, 'Babylon in European Thought', in J. M. Sasson (ed.), *Civilizations of the Ancient Near East* (New York: Charles Scribner's Sons), 67–80.

Lynn, W. T., 1889, 'The Chaldæan Saros', *The Observatory* 12, 261–262.

Lynn, W. T., 1905, 'Richard Dunthorne', *The Observatory* 28, 215–216.

McConnell, A., 2004, 'George Costard (*bap.* 1710, *d.* 1782)', *Oxford Dictionary of National Biography* (Oxford: Oxford University Press).

MacPike, E. F., 1932, *Correspondence and Papers of Edmond Halley* (Oxford: Clarendon Press).

MacPike, E. F., 1937, *Hevelius, Flamsteed and Halley: Three Contemporary Astronomers and Their Mutual Relations* (London: Taylor and Francis).

Marshall, A., 2008, 'The Myth of Scriblerus', *Journal for Eighteenth-Century Studies* 31, 77–99.

Mercier, R., 1994, 'English Orientalists and Mathematical Astronomy', in G. A. Russell (ed.), *The 'Arabick' Interest of the Natural Philosophers in Seventeenth-Century England* (Leiden, New York, Köln: Brill), 158–214.

Minkowski, H., 1991, *Vermutungen über den Turm zu Babel* (Freren: Luca Verlag).

Monod-Cassidy, H., 1980, *Jerome Lalande: Journal d'un Voyage en Angleterre 1763*, Studies on Voltaire and the Eighteenth Century 184 (Oxford: The Voltaire Foundation).

de Montluzin, E. L., 2001, 'Topographical, Antiquarian, Astronomical, and Meteorological Contributions by George Smith of Wigton in the *Gentlesman's Magazine*, 1735–59', *American Notes and Queries* 14:2, 5–12.

de Montluzin, E. L., 2004, 'George Smith of Wigton: *Gentleman's Magazine* Contributor, Unheralded Scientific Polymath, and Shaper of the Aesthetic of the Romantic Sublime', *Eighteenth-Cerntury Life* 28:3, 66–89.

Morando, B., 1995, 'Three Centuries of Lunar and Planetary Ephemerides and Tables', in R. Taton and C. Wilson (eds.), *The General History of Astronomy. Volume 2: Planetary Astronomy from the Renaissance to the Rise of Astrophysics. Part B: The Eighteenth and Nineteenth Centuries* (Cambridge: Cambridge University Press), 251–259.

Mullinger, J. B., 1901, *St John's College* (London: F. E. Robinson & Co.).

Neugebauer, O., 1957, *The Exact Sciences in Antiquity* (Providence: Brown University Press).

Nevill, E. 1906, 'The Early Eclipses of the Sun and Moon', *Monthly Notices of the Royal Astronomical Society* 67, 2–13.

Newton, R. R., 1969, 'Secular Accelerations of the Earth and Moon', *Science* 166, 825–831.

Newton, R. R., 1970, *Ancient Astronomical Observations and the Accelerations of the Sun and Moon* (Baltimore: Johns Hopkins University Press).

Newton, R. R., 1977, *The Crime of Claudius Ptolemy* (Baltimore: Johns Hopkins University Press).

Nichols, J., 1812, *Literary Anecdotes of the Eighteenth Century* (6 volumes, London: Nichols, Son and Bentley).

Ooghe, B., 2007, 'The Rediscovery of Babylonia: European Travellers and the Development of Knowledge on Lower Mesopotamia, Sixteenth to Early Nineteenth Century', *Journal of the Royal Asiatic Society*, Series 3 17, 231–252.

Oppolzer, T. von, 1881, *Syzygientafeln für den Mond nebst aüsfuhrlichen Anwersun zum Gebrauch derselben* (Leipzip: Der astronomischen Gesellschaft).

Pallis, S. A., 1956, *The Antiquity of Iraq. A Handbook of Assyriology* (Copengagen: Munksgaard).

Pedersen, O., 1974, *A Survey of the Almagest* (Odense: Odense University Press).

Peiffer, J., 2002, 'France', in J. W. Dauben and C. J. Scriba (eds.), *Writing the History of Mathematics: Its Historical Development* (Basel: Birkhäuser), 3–43.

Plummer, H. C., 1940–41, 'Jeremiah Horrocks and his *Opera Posthuma*', *Notes and Records of the Royal Society* 3, 39–52.

Popper, N., 2006, '"Abraham, Planter of Mathematics": Histories of Mathematics and Astrology in Early Modern Europe', *Journal of the History of Ideas* 67, 87–106.

Ragep, J., 2010, 'Islamic Reactions to Ptolemy's Imprecisions', in A. Jones (ed), *Ptolemy in Perspective: Use and Criticism of his Work from Antiquity to the Nineteenth Century* (Dordrecht: Springer), 121–134.

Reade, J. E., 1999, 'Early British Excavations at Babylon', in J. Renger (ed.), *Babylon: Focus Mesopotamischer Geschichte, Wiege Früher Gelehrsamkeit, Mythos in der Moderne* (Berlin: Deutsche Orient-Gesellschaft), 47–65.

Reade, J. E., 2008, 'Disappearance and Rediscover', in I. L. Finkel and M. J. Seymour (eds.), *Babylon: Myth and Reality* (London: The British Museum), 13–32.

Rome, A., 1950, 'The Calculation of an Eclipse of the Sun According to Theon of Alexandria', in *Proceedings of the International Congress of Mathematicians* (Providence), 209–219.

Rousseau, G. S., 1987, '"Whicked Whiston" and the Scriblerians: Another Ancients-Moderns Controversy', *Studies in Eighteenth-Century Culture* 17, 17–44.

Russell, G. A., 1993, 'Introduction: The Seventeenth Century: The Age of "Arabic"', in G. A. Russell (ed.), *The 'Arabick' Interest of the Natural Philosophers in Seventeenth-Century England* (Leiden, New York, Köln: Brill), 1–19.

Rutkin, H. D., 2010, 'The Use and Abuse of Ptolemy's *Tetrabiblos* in Renaissance and Early Modern Europe: Two Case Studies (Giovanni Pico della Mirandola and Filippo Fantoni)', in A. Jones (ed), *Ptolemy in Perspective: Use and Criticism of his Work from Antiquity to the Nineteenth Century* (Dordrecht: Springer), 135–149.

Said, S. S., and Stephenson, F. R., 1996, 'Solar and Lunar Eclipse Measurements by Medieval Muslim Astronomers, I: Background', *Journal for the History of Astronomy* 27, 259–273.

Said, S. S., and Stephenson, F. R., 1997, 'Solar and Lunar Eclipse Measurements by Medieval Muslim Astronomers, II: Observations', *Journal for the History of Astronomy* 28, 29–48.

Sarton, G., 1936, 'Montucla (1725–1799): His Life and Works', *Osiris* 1, 519–567.

Schaffer, S., 1977, 'Halley's Atheism and the End of the World', *Notes and Records of the Royal Society of London* 32, 17–40.

Seymour, M., 2008, 'The Tower of Babel in Art', in I. L. Finkel and M. J. Seymour (eds.), *Babylon: Myth and Reality* (London: The British Museum), 132–141.

Sivin, N., 2009, *Granting the Seasons: The Chinese Astronomical Reform of 1280, With a Study of Its Many Dimensions and an Annotated Translation of Its Records* (New York: Springer).

Sivin, N., 2011, 'Mathematical Astronomy and the Chinese Calendar', in J. M. Steele (ed.), *Calendars and Years II: Astronomy and Time in the Ancient and Medieval World* (Oxford: Oxbow Books), 39–51.

Skempton, A. W., 2002, 'Richard Dunthorne', in A. W. Skempton (ed.), *Biographical Dictionary of Civil Engineers. Volume 1 – 1500 to 1830* (London: Thomas Telford), 196–198.

Smith, A., 2008, 'Computerised Resources for Historical Research: Calendars, Chronology and the Life of Jesus Christ', in J. Carvalha (ed.), *Bridging the Gaps: Sources, Methodology and Approaches to Religion in History* (Piza: Piza University Press), 29–56.

Steele, J. M., 2000, *Observations and Predictions of Eclipse Times by Early Astronomers* (Dordrecht: Kluwer).

Steele, J. M, 2003, 'The Use and Abuse of Astronomy in Establishing Absolute Chronologies', *Physics in Canada*, 59 (2003), 243–248.

Steele, J. M., 2004, 'Applied Historical Astronomy: An Historical Perspective', *Journal for the History of Astronomy* 35, 337–355.

Steele, J. M., 2005, 'Ptolemy, Babylon and the Rotation of the Earth', *Astronomy and Geophysics* 46/5, 11–15.

Steele, J. M., 2011, 'Visual Aspects of the Transmission of Babylonian Astronomy and its Reception into Greek Astronomy', *Annals of Science* 68, 543–465.

Steele, J. M., 2012, 'The 'Astronomical Fragments' of Berossos in Context', in J. Haubold, G. Lanfranci, R. Rollinger and J. M. Steele (eds.), *The World of Berossos* (Wiesbaden: Harrasowitz), in press.

Steele, J. M., and Stephenson, F. R., 1998, 'Eclipse Observations made by Regiomontanus and Walther', *Journal for the History of Astronomy* 29, 331–344.

Stephenson, F. R., 1997, *Historical Eclipses and Earth's Rotation* (Cambridge: Cambridge University Press).

Stephenson, F. R., and Clark, D., 1978, *Applications of Early Astronomical Records* (Bristol: Adam Hilger Ltd).

Stephenson, F. R., and Yau, K. K. C., 1992, 'Astronomical Records in the *Ch'un-Ch'iu* Chronicle', *Journal for the History of Astronomy* 23, 31–51.

Stückelberger, A., and Graßhoff, G., (eds.), 2006, *Ptolemaios. Handbuch der Geographie*, 2 vols. (Basel: Schwabe).

Sturdy, D. J., 1995, *Science and Social Status: The Members of the Académie des Sciences, 1666–1750* (Woodbridge: The Boydell Press).

Swerdlow, N. M., 1993, 'Montucla's Legacy: The History of the Exact Sciences', *Journal for the History of Ideas* 54, 299–328.

Swedlow, N. M., 2010, 'Tycho, Longomontanus, and Kepler on Ptolemy's Solar Observations and Theory, Precession of the Equinoxes, and Obliquity of the Ecliptic', in A. Jones (ed.), *Ptolemy in Perspective: Use and Criticism of his Work from Antiquity to the Nineteenth Century* (Dordrecht: Springer), 151–202.

Toomer, G. J., 1984, *Ptolemy's Almagest* (London: Duckworth).

Toomer, G. J., 1996, *Eastern Wisedome and Learning. The Study of Arabic in Seventeeth-Century England* (Oxford: Clarendon Press).

Turnbull, H. W., 1960, *The Correspondence of Isaac Newton*, vol. 2 (Cambridge: Cambridge University Press).

van der Waerden, B. L., 1958, 'Drei umstrittene Mondfinsternisse bei Ptolemaios', *Museum Helveticum* 15, 106–106.

Waff, C. B., 1977, 'Newton and the Motion of the Moon: An Essay Review', *Centaurus* 21, 64–75.

Wegener, U. B., 1995, *Die Faszination des Maßlosen. Der Turmbau zu Babel von Pieter Breugel bis Athanasius Kircher* (Hildesheim–Zürich–New York: Olms).

Wepster, S., 2010, *Between Theory and Observations: Tobias Mayer's Explorations of Lunar Motion, 1751–1755* (New York: Springer).

Whiteside, D. T., 1976, 'Newton's Lunar Theory: From High Hope to Disenchantment', *Vistas in Astronomy* 19, 1–21.

Wilson, C., 1980, 'Perturbations and Solar Tables from Lacaille to Delambre: the Rapprochement of Observation and Theory, Part I', *Archive for History of Exact Sciences* 22, 53–304.

Wilson, C., 1985, 'The Great Inequality of Jupiter and Saturn: From Kepler to Laplace', *Archive for History of Exact Sciences* 33, 15–290.

Wilson, C., 1987, 'On the Origin of Horrocks's Lunar Theory', *Journal for the History of Astronomy* 18, 77–94.

Wilson, C., 1989a, 'Predictive Astronomy in the Century After Kepler', in R. Taton and C. Wilson (eds.), *The General History of Astronomy. Volume 2: Planetary Astronomy from the Renaissance to the Rise of Astrophysics. Part A: Tycho Brahe to Newton* (Cambridge: Cambridge University Press), 159–160.

Wilson, C., 1989b, 'The Newtonian Achievement in Astronomy', in R. Taton and C. Wilson (eds.), *The General History of Astronomy. Volume 2: Planetary Astronomy from the Renaissance to the Rise of Astrophysics. Part A: Tycho Brahe to Newton* (Cambridge: Cambridge University Press), 233–274.

Wilson, C., 1995, 'The Problem of the Perturbation Analytically Treated: Euler, Clairaut, d'Alembert', in R. Taton and C. Wilson (eds.), *The General History of Astronomy. Volume 2: Planetary Astronomy from the Renaissance to the Rise of Astrophysics. Part B: The Eighteenth and Nineteenth Centuries* (Cambridge: Cambridge University Press), 89–107.

Wilson, C., 2010, *The Hill-Brown Theory of the Moon's Motion: Its Coming-to-be and Short-lived Ascendancy (1877–1984)* (New York: Springer).

Winstanley, D. A., 1935, *Unreformed Cambridge* (Cambridge: Cambridge University Press).

Young, B. W., 2004, 'Heathcote, Ralph (1721–1795)', *Oxford Dictionary of National Biography* (Oxford: Oxford University Press).

Zhmud, L., 2006, *The Origin of the History of Science in Classical Antiquity* (Berlin–New York: Walter de Gruyter).

Index

Printed in the United States
By Bookmasters